Comparative Innovation Policy

INNOVATIVE FLANDERS

Innovation Policies for the 21st Century

Report of a Symposium

Committee on Comparative Innovation Policy:
Best Practice for the 21st Century

Board on Science, Technology, and Economic Policy

Policy and Global Affairs

Charles W. Wessner, Editor

NATIONAL RESEARCH COUNCIL
OF THE NATIONAL ACADEMIES

THE NATIONAL ACADEMIES PRESS
Washington, D.C.
www.nap.edu

THE NATIONAL ACADEMIES PRESS 500 Fifth Street, N.W. Washington, DC 20001

NOTICE: The project that is the subject of this report was approved by the Governing Board of the National Research Council, whose members are drawn from the councils of the National Academy of Sciences, the National Academy of Engineering, and the Institute of Medicine. The members of the committee responsible for the report were chosen for their special competences and with regard for appropriate balance.

This study was supported by: Contract/Grant No. SB1341-03-C-0032 between the National Academy of Sciences and the U.S. Department of Commerce; Contract/Grant No. OFED-381989 between the National Academy of Sciences and Sandia National Laboratories; and Contract/Grant No. NAVY-N00014-05-G-0288, DO #2, between the National Academy of Sciences and the Office of Naval Research. This material is based upon work also supported by the Defense Advanced Research Projects Agency Defense Sciences Office, DARPA Order No. K885/00, Program Title: Materials Research and Development Studies, Issued by DARPA/CMD under Contract #MDA972-01-D-0001. Additional funding was provided by Intel Corporation, International Business Machines, the Palo Alto Research Center, and Google. Any opinions, findings, conclusions, or recommendations expressed in this publication are those of the author(s) and do not necessarily reflect the views of the organizations or agencies that provided support for the project.

International Standard Book Number-13: 978-0-309-11606-0
International Standard Book Number-10: 0-309-11606-6

Limited copies are available from Board on Science, Technology, and Economic Policy, National Research Council, 500 Fifth Street, N.W., W547, Washington, DC 20001; 202-334-2200.

Additional copies of this report are available from the National Academies Press, 500 Fifth Street, N.W., Lockbox 285, Washington, DC 20055; (800) 624-6242 or (202) 334-3313 (in the Washington metropolitan area); Internet, http://www.nap.edu.

Printed in the United States of America

Committee on Comparative Innovation Policy: Best Practice for the 21st Century*

William J. Spencer, *Chair (through August 2007)*
Chairman Emeritus, *retired*
SEMATECH

Kenneth S. Flamm, *Vice Chair*
Dean Rusk Chair in International Affairs
Lyndon B. Johnson School of Public Affairs
University of Texas at Austin
and STEP Board

Alice H. Amsden
Professor of Political Economy
Massachusetts Institute of Technology

Gail H. Cassell
Vice President, Scientific Affairs and
Distinguished Lilly Research Scholar for Infectious Diseases
Eli Lilly and Company

Carl J. Dahlman
Henry R. Luce Associate Professor
Edmund A. Walsh School of Foreign Service
Georgetown University

Alan Wm. Wolff, *Chair (August 2007-present)*
Partner
Dewey & LeBoeuf LLP
and STEP Board

Mary L. Good, *Vice Chair*
Donaghey University Professor
Dean, Donaghey College of Information Science and Systems Engineering
University of Arkansas at Little Rock
and STEP Board

Lewis S. Edelheit
Senior Research and Technology Advisor, *retired*
General Electric

Bronwyn Hall
Professor of Economics
University of California at Berkeley

Mark B. Myers
Senior Vice President, *retired*
Xerox

*As of May 2008.

Project Staff*

Charles W. Wessner
Study Director

Sujai J. Shivakumar
Senior Program Officer

Alan Anderson
Consultant

David E. Dierksheide
Program Officer

Jeffrey C. McCullough
Program Associate

*As of May 2008.

For the National Research Council (NRC), this project was overseen by the Board on Science, Technology, and Economic Policy (STEP), a standing board of the NRC established by the National Academies of Sciences and Engineering and the Institute of Medicine in 1991. The mandate of the STEP Board is to integrate understanding of scientific, technological, and economic elements in the formulation of national policies to promote the economic well-being of the United States. A distinctive characteristic of STEP's approach is its frequent interactions with public- and private-sector decision makers. STEP bridges the disciplines of business management, engineering, economics, and the social sciences to bring diverse expertise to bear on pressing public policy questions. The members of the STEP Board* and the NRC staff are listed below:

Lawrence S. Summers, *Chair*
Charles W. Eliot Professor
Kennedy School of Government
Harvard University

Timothy F. Bresnahan
Landau Professor in Technology and the Economy
Stanford University

Lewis W. Coleman
President
DreamWorks Animation

Kenneth S. Flamm
Dean Rusk Chair in International Affairs
Lyndon B. Johnson School of Public Affairs
University of Texas at Austin

Ralph E. Gomory
President Emeritus
Alfred P. Sloan Foundation

Mary L. Good
Donaghey University Professor
Dean, Donaghey College of Information Science and Systems Engineering
University of Arkansas at Little Rock

Edward E. Penhoet, *Vice Chair*
Director
Alta Partners

Amory Houghton, Jr.
Former Member of Congress

David T. Morgenthaler
Founding Partner
Morgenthaler Ventures

Joseph P. Newhouse
John D. MacArthur Professor of Health Policy and Management
Director, Division of Health Policy Research and Education
Harvard University

Arati Prabhakar
General Partner
U.S. Venture Partners

William J. Raduchel
Chairman
Opera Software ASA

Jack W. Schuler
Chairman
Ventana Medical Systems, Inc.

Alan Wm. Wolff
Partner
Dewey & LeBoeuf LLP

---

*As of May 2008.

STEP Staff*

Stephen A. Merrill
Executive Director

Sarah R. Carter
Christine Mirzayan
Science & Technology
Policy Fellow

David E. Dierksheide
Program Officer

Charles W. Wessner
Program Director

Jeffrey C. McCullough
Program Associate

Sujai J. Shivakumar
Senior Program Officer

Mahendra Shunmoogam
Program Associate

*As of May 2008.

Contents

Affiliations as of September 2006.

Affiliations as of September 2006.

Affiliations as of September 2006.

Affiliations as of September 2006.

Preface

Recognizing that a capacity to innovate and commercialize new high-technology products is increasingly a part of the international competition for economic leadership, governments around the world are taking active steps to strengthen their national innovation systems. These steps underscore the belief that the rising costs and risks associated with new potentially high-payoff technologies, and the growing global dispersal of technical expertise, require national research and development programs to support new and existing high-technology firms within their borders.

What is the impact of this new international competition for the United States? In a recent report, the National Academies warned that "this nation must prepare with great urgency to preserve its strategic and economic security," adding that "the United States must compete by optimizing its knowledge-based resources, particularly in science and technology, and by sustaining the most fertile environment for new and revitalized industries and the well-paying jobs they bring."[1]

Responding to this challenge requires that we recognize that the nature and terms of economic competition are shifting.[2] U.S. policymakers need to be aware

---

[1]National Academy of Sciences/National Academy of Engineering/Institute of Medicine, *Rising Above the Gathering Strom: Energizing and Employing America for a Brighter Future,* Washington, D.C.: The National Academies Press, 2007.

[2]Kent Hughes has argued in this regard that the challenges of the 21st century require new strategies that take account of new technologies, new global competitors, as well as new national priorities concerning national security and the environment. See Kent Hughes, *Building the Next American Century: The Past and Future of American Economic Competitiveness,* Washington, D.C.: Woodrow Wilson Center Press, 2005, Chapter 14.

of the wide variety of innovation and competitiveness policies that many nations have adopted. These policies are designed to build research capacities and to acquire knowledge, and then to transition that knowledge directly to companies and support their development.

Some nations have developed well-financed and integrated national programs that are designed to shift the terms of international competition. Other national programs, while more modest in scale, provide essentially market-based incentives to encourage the transition of new technologies to the market. Yet, even these can have a significant impact on the terms of competition. While institutions and the scale of funding vary across the globe, a comparative perspective is necessary to help us understand what policies are succeeding and why, how selected policies might be successfully adapted in the U.S. context, and what existing U.S. programs might be enhanced.

With these objectives in mind, the National Research Council's Board on Science, Technology, and Economic Policy (STEP) has embarked on a study of selected foreign innovation programs in comparison with major U.S. programs. Recognizing the importance of targeted government promotional policies relative to innovation, the analysis, carried out under the direction of an ad hoc Committee, is to include a review of the goals, concept, structure, operation, funding levels, and evaluation of foreign programs designed to advance the innovation capacity of national economies and enhance their international competitiveness.

Definitions of Innovation and Competitiveness

We define innovation as the transformation of an idea into a marketable product or service, a new or improved manufacturing or distribution process, or even a new method of providing a social service. This transformation involves an adaptive network of institutions that encompass a variety of informal and formal rules, norms, and procedures—a national innovation ecosystem—that shape how individuals and corporate entities create knowledge and collaborate to bring new products and services to market.

If we define competitiveness as the ability to gain market share by adding value better than others in the globalized economic environment, the ability of these actors to collaborate successfully within a given innovation ecosystem gains significance. Recognizing this, policymakers around the world are supporting a variety of initiatives to reinforce their national innovation ecosystems as a way of improving their national competitiveness.

THE CONTEXT OF THIS REPORT

Since 1991 the STEP Board has undertaken a program of activities to improve policymakers' understanding of the interconnections among science, technology, and economic policy and their importance to the American economy and its international competitive position. The Board's interest in comparative innovation policies derives directly from its mandate.

This mandate has previously been reflected in STEP's widely cited volume, *U.S. Industry in 2000,* which assesses the determinants of competitive performance in a wide range of manufacturing and service industries, including those relating to information technology.[3] The Board also undertook a major study, chaired by Gordon Moore of Intel, on how government-industry partnerships can support the growth and commercialization of productivity enhancing technologies.[4] Reflecting a growing recognition of the importance of the surge in productivity since 1995, the Board also launched a multifaceted assessment, exploring the sources of growth, measurement challenges, and the policy framework required to sustain the New Economy.[5]

The current study on Comparative Innovation Policy builds on STEP's experience to develop an international comparative analysis focused on U.S. and foreign innovation programs. To open this analysis, the Committee held a symposium on April 15, 2005, which drew together leading academics, policy analysts, and senior policymakers from around the globe to describe their national innovation programs and policies, outline their objectives, and highlight their achievements.[6] Follow-up symposia in Taipei and Tokyo in January 2006 focused on the evolution of the Taiwanese and Japanese innovation systems over the past decade. The Committee also convened a major symposium in Washington in June 2006 that identified current trends in the Indian innovation system and the new U.S.-India innovation partnership.[7]

---

[3]National Research Council, *U.S. Industry in 2000: Studies in Competitive Performance,* David C. Mowery, ed., Washington, D.C.: National Academy Press, 1999.

[4]This summary of a multivolume study provides the Moore Committee's analysis of best practices among key U.S. public-private partnerships. See National Research Council, *Government-Industry Partnerships for the Development of New Technologies: Summary Report,* Charles W. Wessner, ed., Washington, D.C.: The National Academies Press, 2003. For a list of U.S. partnership programs, see Christopher Coburn and Dan Berglund, *Partnerships: A Compendium of State and Federal Cooperative Programs,* Columbus, OH: Battelle Press, 1995.

[5]National Research Council, *Enhancing Productivity Growth in the Information Age: Measuring and Sustaining the New Economy,* Dale W. Jorgenson and Charles W. Wessner, eds., Washington, D.C.: The National Academies Press, 2007.

[6]For a summary of this conference, see National Research Council, *Innovation Policies for the 21st Century,* Charles W. Wessner, ed., Washington, D.C.: The National Academies Press, 2007.

[7]For a summary of this conference, see National Research Council, *India's Changing Innovation System,* Charles W. Wessner and Sujai J. Shivakumar, eds., Washington, D.C.: The National Academies Press, 2007.

In September 2006, the Committee held a major international symposium on "Synergies in Regional and National Innovation Policies in the Global Economy" in Flanders, Belgium. This event reviewed European Union, national, and regional innovation policies in Europe. The Committee met with representatives from policymakers and academics in Leuven, in the Flanders region of Belgium, a major university and research center with a strong commercialization record. Leuven is also home to IMEC, one of the leading microelectronics research facilities in the world and the flagship of Flemish technology policy. This report provides a summary of the symposium.

ACKNOWLEDGMENTS

We are grateful for the support of the Defense Advanced Research Projects Agency, the National Institute of Standards and Technology, the National Science Foundation, the Office of Naval Research, and Sandia National Laboratories.

We are especially grateful to our hosts in Flanders for their help in organizing this conference. Foremost among them is the former Minister Fientje Moerman, whose leadership on innovation policies has made a lasting contribution to Flemish growth and competitiveness. We are indebted to our hosts at IMEC, notably Jan Wauters and Imke Debecker. We are also grateful to Peter Spyns, Emmelie Tindemans, Lea De Pauw, and Marleen De Leenheer of the Flemish Department of Economy, Science and Innovation for their hospitality and attention to the many practicalities of the symposium. Bart Hendrickx, the former diplomatic representative of Flanders for the United States deserves special mention. Without his interest and commitment, the symposium would not have occurred.

On the U.S. side, we are indebted to Alan Anderson for his preparation of this meeting summary and to Sujai Shivakumar and Alan Anderson for preparing the Introduction to this volume. Several members of the STEP staff also deserve recognition for their contributions to the preparation of this report, including Jeffrey McCullough and David Dierksheide for their role in preparing the conference and getting this report ready for publication.

NATIONAL RESEARCH COUNCIL REVIEW

This report has been reviewed in draft form by individuals chosen for their diverse perspectives and technical expertise, in accordance with procedures approved by the National Academies' Report Review Committee. The purpose of this independent review is to provide candid and critical comments that will assist the institution in making its published report as sound as possible and to ensure that the report meets institutional standards for quality and objectivity. The review comments and draft manuscript remain confidential to protect the integrity of the process.

We wish to thank the following individuals for their review of this report: Daryl Hatano, Semiconductor Industry Association; Göran Marklund, VINNOVA; Luc Soete, Maastricht University; and Peter Spyns, Department of Economy, Science and Innovation, Flemish Government.

Although the reviewers listed above have provided many constructive comments and suggestions, they were not asked to endorse the content of the report, nor did they see the final draft before its release. Responsibility for the final content of this report rests entirely with the authors and the institution.

Alan Wm. Wolff Charles W. Wessner

I

INTRODUCTION

Innovative Flanders: Innovation Policies for the Twenty-first Century

Recognizing that innovation is the key to international competitiveness in the 21st century, policymakers around the world are seeking more effective ways to translate scientific and technological knowledge into new products, processes, and businesses. They have initiated major programs, often with substantial funding, that are designed to attract, nurture, and support innovation and high-technology industries within their national economies.

To help U.S. policymakers become more aware of these developments, a committee of the National Academies' Board on Science, Technology, and Economic Policy undertook a review of the goals, concept, structure, operation, funding levels, and evaluation efforts of significant innovation programs around the world. As a part of this effort, the committee identified Flanders, a region of Belgium with substantial autonomy, which is recognized for its comprehensive approach to innovation. Based on initial meetings in Washington and Brussels, and with the endorsement of Flanders Vice Minister-President Fientje Moerman, it was agreed to organize a conference that would review regional innovation policies in the context of the policies and programs of the Flanders government, and their interaction with those of the European Union.[1]

This chapter highlights the main points of this conference. It begins with an overview of the changing landscape of global innovation and reviews the role

---

[1]Mrs. Moerman resigned as Flanders' Vice Minister-President and Minister of Economy, Enterprises, Innovation, Science, and Foreign Trade in October 2007. The conference reported here was held in September 2006. Titles and positions of all the participants reported in this volume are those of the date of the conference.

Box A
What is Innovation?

"Innovation is a strategy that provides resources to talented people in an atmosphere which promotes creativity and is focused on outcomes ranging from new products, to customer satisfaction, to new scientific insights, to improved processes, to improved social programs. [It is] designed to create wealth and/or improve the human condition."

Dr. Mary Good, University of Arkansas at Little Rock

that public-private partnerships play in advancing American competitiveness. The chapter then turns to review the initiatives taken by the Flanders government to reinforce its position further as a global center of research and innovation.

While the American and Flanders economies differ vastly in scale and structure, both confront common challenges in innovation, including the need to transform existing institutions and invent new policies mechanisms for the future. A premise of the conference—and hence this report—is that a comparative perspective is necessary to understand the global environment for innovation-based competition. A detailed summary of the insights, observations, and the status of current policies captured in the conference proceedings can be found in the next chapter.

THE GLOBALIZATION OF INNOVATION

Since the Second World War, the high standard of living found in the United States and Western Europe has been built on competitive markets that reward the innovator while providing consumers with better and more affordable products. In the United States, this potential for innovation has been sustained by a culture of entrepreneurship that encourages risk-taking by providing substantial rewards buttressed by robust government funding for basic science and technology, and reinforced by an open research and development (R&D) system that attracts the best minds from around the world.[2]

While still a powerful model, this paradigm began to change with the emergence of a distinctly new competitive environment in the 1990s. The introduction of new information and communications technologies across the world and the rise of new low-wage, high-skill entrants on the global stage have altered the

[2]See the presentation by Dr. Mary Good in the Proceedings section of this volume.

Box B
Innovation in Flanders

About the size of Connecticut and with a population of about six million, Flanders encompasses the Dutch speaking region of Belgium. Constitutional reforms in Belgium, begun in the 1970s, now provide the Flanders government with considerable autonomy to pursue its own social and economic policies.

Until the middle of the 20th century, Flanders lagged economically behind Belgium's French speaking region of Wallonia. With the decline of Wallonia's powerful coal and iron industries after the Second World War, more modern business growth came to Flanders. By the end of the 20th century, Flanders was home to dynamic auto assembly, pharmaceuticals, engineering, metal products, food processing, chemicals, and brewing industries. Exports of manufactured products accounted for nearly 80 percent of Flanders' gross domestic product (GDP).

Recognizing the need to secure its peoples' future prosperity in a rapidly changing and competitive global environment, the Flanders government decided to strengthen its own high-technology base, and has since implemented a broad range of programs to enhance its innovation capacity—the focus of this volume.

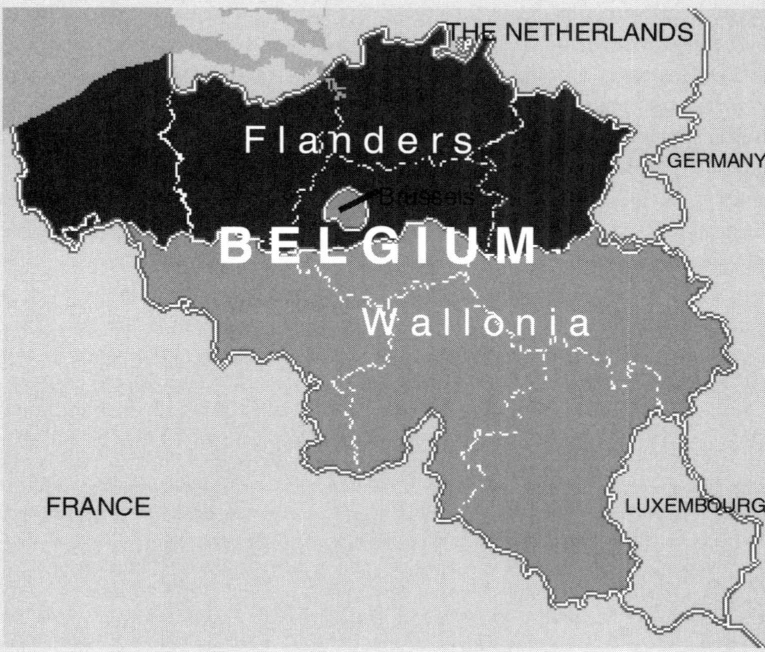

FIGURE B-1 Map of Belgium.

landscape of innovation, creating new challenges for the continued technological leadership of the United States.[3]

Technological Transformations

The information and communications revolution, made possible in large part by faster and cheaper semiconductor products, has changed the economics of innovation. Taking advantage of the potential offered by new information and communications technologies, many large firms have transformed themselves from vertically integrated enterprises, often with significant in-house R&D capabilities, into flat, virtual, and globally networked enterprises.[4] With Moore's Law, which predicts the regular and rapid increase in microprocessor capacity, expected to continue for at least another 15 years, the continued decline in cost and increase in capacity of information technologies is likely to continue to underpin this revolution.[5]

In this new paradigm, talent does not necessarily have to be based in or drawn to the United States, but can be accessed from across the globe. As Mark Myers of the University of Pennsylvania noted in his conference remarks, large firms no longer invest in in-house scientific research as they once did, drawing instead on needed technologies through investment, partnerships, and acquisitions of small innovative firms. Production and sales are similarly fragmented, based on worldwide supply chains and a worldwide customer base.

As Dr. Myers noted, this new reality means that each nation must have policies that address the globalization dynamic. New models of cooperation among governments, industries, universities, and others are necessary to sustain the "knowledge commons" on which innovation depends. And new types of investments in education are necessary to prepare the workforce of the future even as skilled workers migrate with increasing ease across the world.

---

[3]These concerns are highlighted in a recent report of the National Academies. See National Academy of Sciences, National Academy of Engineering, Institute of Medicine, *Rising Above the Gathering Storm: Energizing and Employing America for a Brighter Economic Future,* Washington, D.C.: The National Academies Press, 2007. Drawing attention to the possibility of an abrupt loss of U.S. leadership in science and innovation, this report led to the passage of the *America Competes Act of 2007.* This act, passed with bipartisan support in Congress, focuses on increasing research investment, strengthening educational opportunities in science, technology, engineering, and mathematics from elementary through graduate school, and developing the nation's innovation infrastructure.

[4]For a review of some of the implications of the ongoing revolution in information and communications technology for businesses, see William J. Raduchel, "The End of Stovepiping," in National Research Council, *The Telecommunications Challenge: Changing Technologies and Evolving Policies,* Charles W. Wessner, ed., Washington, D.C.: The National Academies Press, 2006, p. 31.

[5]For an analysis of the nature of Moore's Law and its impact on the U.S. productivity growth, see National Research Council, *Enhancing Productivity Growth in the Information Age,* Dale W. Jorgenson and Charles W. Wessner, eds., Washington, D.C.: The National Academies Press, 2007.

Box C
A Chinese Perspective on Innovation and National Competitiveness

"In today's world, the core of each country's competitive strength is intellectual innovation, technological innovation, and high-tech industrialization."

President Jiang Zemin
August 23, 1999

The Rise of New Entrants

Another challenge to continuing U.S. leadership in innovation and competitiveness comes from newly competitive participants in the global economy. China, most notably, combines the advantages of high-skill and low-wage knowledge workers with substantial state and foreign investments backed by a strong sense of national purpose in acquiring new capabilities and participating in product markets based on advanced technologies.[6]

One element of this strategy focuses on attracting and developing high-technology industries to the Mainland. As Alan Wolff of Dewey Ballantine LLP noted at the conference, China's leaders see the acquisition of technological capabilities and control of national market as a means of maintaining national autonomy and generating political and military strength. (See Box C.)

This high-level commitment is evident in the rapid rise in Chinese R&D expenditure. In 1999, China's R&D spending accounted for 6 percent of the total world expenditures in R&D. By 2005, China accounted for 13 percent of the world total of $836 billion spent on R&D.[7] Mr. Wolff reported that China plans to increase its R&D spending to 2.5 percent of GDP by 2010, raising it to international target levels.

In addition to national focus and generous funding, China has also adopted powerful policies to encourage innovation. These policies include exemptions from sales tax income earned from the transfer of technology developed exclusively through foreign direct investment in R&D, a 50 percent discount in corporate income tax for foreign R&D investors with rising development expenses,

[6]For a comprehensive review of the innovation policies of India, another major new entrant, see National Research Council, *India's Changing Innovation System: Achievements, Challenges, and Opportunities for Cooperation,* Charles W. Wessner and Sujai J. Shivakumar, eds., Washington, D.C.: The National Academies Press, 2007.

[7]Based on purchasing power parity. See Organisation for Economic Co-operation and Development, *Main Science and Technology Indicators,* Paris: Organisation for Economic Co-operation and Development, 2006.

and (like many countries) procurement regulations that favor national producers. The central and regional governments are also spending substantial sums to support leading industries, such as the construction of advanced semiconductor fabrication facilities.[8] Like governments elsewhere, albeit on a larger scale, the public authorities have set aside large tracts of land for information technology and biotechnology science parks, and are providing incentives for major U.S. and European firms to conduct research and development in China. While some of these efforts, particularly those involving a "top-down" approach, may face drawbacks from bureaucratic rigidities, the sheer scale of China's efforts will continue to have a global impact.

THE INNOVATION CHALLENGE FOR THE UNITED STATES

The emergence of China as a rapidly growing economy offers major growth opportunities for U.S. firms just as China's desire to acquire and develop advanced technology poses significant challenges for U.S. policymakers. In any event, for the United States to maintain its leadership as an innovative economy, it has to adapt its policies to address these new technological and competitive realities.

Dr. Mary Good of the University of Arkansas underscored the nature of the challenge faced by the United States. These challenges include changing demographics and unfavorable trends in investments on science. Noting that over a third of the science and technology (S&T) graduate students in the United States are foreign-born, and that nearly 60 percent of engineering graduates are foreign-born, she said that U.S. innovation depends on the availability and continued presence of these foreign-born students.[9] The question is whether they will continue to come and stay as other countries quickly build up their own research universities and job opportunities and our own immigration system discourages them from staying. What is needed, she affirmed, are immigration policies that admit educated newcomers while restricting illegal immigrants.

At the same time, Dr. Good noted, the United States is not investing sufficiently in its future innovation capacity. Funding for public universities has declined, making it more difficult to replace retiring generations of scientists and

---

[8]For a discussion of policies adopted by the People's Republic of China to support its semiconductor industry, see Thomas R. Howell, "New Paradigms for Partnerships: China Grows a Semiconductor Industry," in National Research Council, *Innovation Policies for the 21st Century,* Charles W. Wessner, ed., Washington, D.C.: The National Academies Press, 2007.

[9]For related analysis, see National Research Council, *Policy Implications of International Graduate Students and Postdoctoral Scholars in the United States,* Washington, D.C.: The National Academies Press, 2005. In 2003, international students earned 38 percent of the U.S.-awarded S&E doctorates and 58.9 percent of the engineering doctorates. See National Science Foundation, *Science and Engineering Doctorate Awards: 2003,* NSF 05-300, Arlington, VA: National Science Foundation, 2004. Data are available at <*http://www.nsf.gov/sbe/srs/nsf05300/tables/tab3.xls*>.

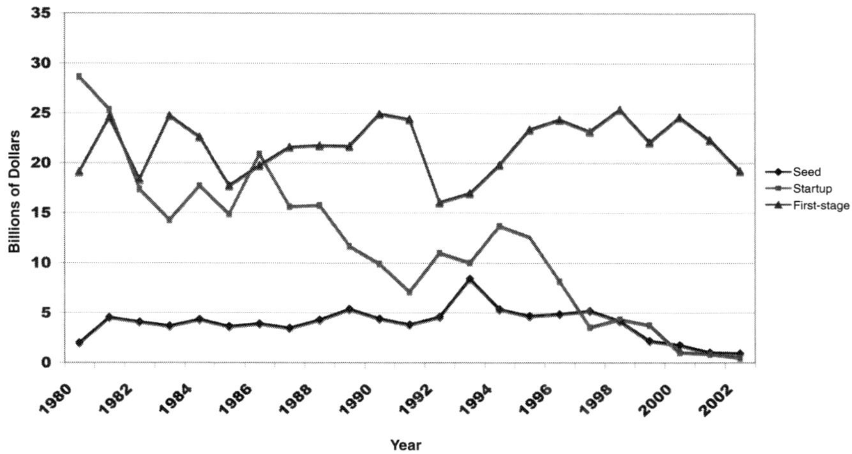

FIGURE 1 The collapse of the U.S. seed and first-stage venture capital funding: dwindling high-risk investments.
SOURCE: National Science Board, *Science and Engineering Indicators 2004*, Arlington VA: National Science Foundation, 2004.

engineers.[10] Another factor weighing on the U.S. innovation system is declining investment in R&D from a variety of sources.[11] While the federal investment has risen in constant dollars since 1976, almost all this increase has gone to the defense sector—where the focus is on development rather than on path-breaking research. Likewise, R&D spending by business is also characterized by a focus on development over research. Dr. Good pointed out that this focus on later stage development is also reflected in venture capital funding, where early-stage funding for small R&D firms "is fast disappearing, and that's got to change." (See Figure 1.)

According to Dr. Good, sustaining America's innovative capacity requires that state and national policymakers pay attention to a set of three interlocking

[10]See Peter R. Orszag and Thomas J. Kane, "Funding Restrictions at Public Universities: Effects and Policy Implications," *Brookings Institution Working Paper,* September 2003. The authors note that public educational spending per full-time equivalent student has declined at public institutions relative to private institutions, from about 70 percent in 1977 to about 58 percent in 1996. Since roughly three-quarters of college students are enrolled at public institutions, they note that any decline in the quality of the nation's public universities could have troubling implications. At the same time, they acknowledge that reductions in spending need not translate into a proportional reduction in quality.

[11]See Kei Koizumi, "Historical Trends in Federal R&D," *AAAS Report XXXII: Research and Development FY2008,* Chapter 2, AAAS Publication Number 07-1A, Washington, D.C.: American Association for the Advancement of Science. Access at <*http://www.aaas.org/spp/rd/08pch2.htm*>.

Box D
The Role of Public-Private Partnerships

"Partnerships facilitate the transfer of scientific knowledge to real products; they represent one means to improve the output of the U.S. innovation system. Partnerships help by bringing innovations to the point where private actors can introduce them to the market. Accelerated progress in obtaining the benefits of new products, new processes, and new knowledge into the market has positive consequences for economic growth and human welfare."[a]

Government-Industry Partnerships for the Development of New Technologies
A Report of the National Academies

---

[a]For an analysis of the conditions necessary for successful public-private partnerships, see the findings and recommendations of the NRC Committee on Government-Industry Partnerships, chaired by Gordon Moore. See National Research Council, *Government-Industry Partnerships for the Development of New Technologies: Summary Report*, Charles W. Wessner, ed., Washington, D.C.: The National Academies Press, 2003, pp. 2-3.

priorities—expanding the nation's talent base, investing in R&D of unexplored areas, and building the infrastructure for collaboration needed to bring new ideas to the market. She summarized the challenges faced by the United States as follows:

- How do you get talent that does what you need it to do?
- How do you raise sufficient support to give that talent opportunity?
- How do you create an infrastructure capable of creating new and exciting things?

In answering these questions, several conference participants pointed to the role that public-private partnerships—involving cooperative R&D activities among industry, universities, and government laboratories—can and have played in accelerating innovation in the United States. (See Box D.)

The case of the semiconductor industry, seen next, illustrates how partnerships have contributed directly to furthering the global competitiveness of a leading U.S. industry.

The SEMATECH Research Consortium

In the 1980s, American semiconductor industry leaders, facing growing competition from Japan, became concerned that they needed to improve manufacturing quality and resolved to find a way to improve the situation collectively.[12] Despite the independence and fierce competitiveness among firms in the industry, the Semiconductor Industry Association took the unusual step of approaching the government and making the argument that active collaboration at the pre-competitive stage was necessary for the sake of long-term U.S. economic competitiveness and national security.[13]

SEMATECH, which brought together most of the largest semiconductor companies in the United States, was launched in 1987 as a new experiment in U.S. R&D strategy. This consortium has since been widely credited with playing a significant role in the resurgence of the U.S. semiconductor industry.[14] Its perceived success has stimulated similar cooperative efforts in Japan and Europe—including the Interuniversity Microelectronics Centre (IMEC), a microelectronics research facility on the outskirts of Leuven in Flanders. Today, SEMATECH continues to play a central role in developing the nanotechnologies necessary to move semiconductor research beyond CMOS and into the future.[15]

In his conference presentation, Kenneth Flamm of the University of Texas said that enhanced research collaboration, made possible by SEMATECH, helped to accelerate the rate of innovation in semiconductor technology and contributed to a rapid decline in the price of semiconductors.[16] The development of a semiconductor technology roadmap in particular helped "coordinate the complex process of technology development to a point where products could all come on

---

[12]See Jeffrey T. Macher, David C. Mowery, and David A. Hodges, "Semiconductors," in *U.S. Industry in 2000: Studies in Competitive Performance,* David C. Mowery, ed., Washington, D.C.: National Academy Press, 1999.

[13]For a first-hand account of the formation of the SEMATECH consortium, see Gordon Moore, "The SEMATECH Contribution," in National Research Council, *Securing the Future: Regional and National Programs to Support the Semiconductor Industry,* Charles W. Wessner, ed., Washington, D.C.: The National Academies Press, 2003. For a view from the Semiconductor Industry Association at that time, see also Andrew Procassini, *Competitors in Alliance: Industry Associations, Global Rivalries, and Business-Government Relations,* New York: Greenwood Publishing, 1995.

[14]For an overview of SEMATECH, see National Research Council, *Conflict and Cooperation in National Competition for High-Technology Industry*, Washington, D.C.: National Academy Press, 1996, pp. 148-151. For an analysis of the empirical evidence, see Kenneth Flamm, "SEMATECH Revisited: Assessing Consortium Impacts on Semiconductor Industry R&D," in National Research Council, *Securing the Future: Regional and National Programs to Support the Semiconductor Industry, op. cit.* See also Peter Grindley, David C. Mowery, and Brian Silverman, "SEMATECH and Collaborative Research: Lessons in the Design of High Technology Consortia," *Journal of Policy Analysis and Management,* 13(4):723-758, 1994.

[15]For a review of new product trends and the future research directions in semiconductor technology, see the remarks by George Scalise, President of the Semiconductor Industry Association, in the Proceedings chapter of this volume.

[16]See Kenneth Flamm, "Economic Impacts of SEMATECH on Innovation in Semiconductors" in the Proceedings chapter of this volume.

line when needed to advance manufacturing."[17] This enhanced pre-competitive research collaboration has been a source of strength to the U.S. semiconductor industry. In turn, the industry's continued dynamism and growth is linked to a rise in the long-term growth trajectory of the United States.[18]

The Role of Innovation Awards

As in the case of SEMATECH, innovation award programs were introduced in the 1980s to address concerns about the international competitiveness of the United States. Drawing on a growing body of evidence that small businesses were assuming an increasingly important role in both innovation and job creation, David Birch, a pioneer in entrepreneurship and small business research, and others, suggested that national policies should promote and build on the competitive strength offered by small businesses.[19]

The Small Business Innovation Research (SBIR) program was established in 1982 as a way to channel federal R&D funds to small businesses. The program was designed to take advantage of the R&D expertise that is often unique to small businesses to meet the mission needs of various government agencies.[20] Moreover, as Charles Wessner of the National Research Council noted in his conference presentation, competitively awarded SBIR grants encourage new entrepreneurship. By signaling information about promising new technologies to

[17]For an analysis of the semiconductor roadmap experiment, see William J. Spencer and T. E. Seidel, "International Technology Roadmaps: The U.S. Semiconductor Experience," in National Research Council, *Productivity and Cyclicality in Semiconductors, Trends, Implications, and Questions,* Dale W. Jorgenson and Charles W. Wessner, eds., Washington, D.C.: The National Academies Press, 2004.

[18]See Dale W. Jorgenson and Kevin J. Stiroh, "Raising the Speed Limit: Economic Growth in the Information Age," in National Research Council, *Measuring and Sustaining the New Economy,* Dale W. Jorgenson and Charles W. Wessner, eds., Washington, D.C.: The National Academies Press, 2002. See also National Research Council, *Enhancing Productivity in the Information Age,* op. cit.

[19]David L. Birch, "Who Creates Jobs?" *The Public Interest,* 65:3-14, 1981. Birch's work greatly influenced perceptions of the role of small firms. Over the past 20 years, it has been carefully scrutinized, leading to the discovery of some methodological flaws, namely making dynamic inferences from static comparisons, confusing gross and net job creation, and admitting biases from chosen regression techniques. See S. J. Davis, J. Haltiwanger, and S. Schuh, "Small Business and Job Creation: Dissecting the Myth and Reassessing the Facts," Working Paper No. 4492, Cambridge, MA: National Bureau of Economic Research, 1993. These methodological fallacies, however, "ha[ve] not had a major influence on the empirically based conclusion that small firms are over-represented in job creation," according to Per Davidsson. See Per Davidsson, "Methodological Concerns in the Estimation of Job Creation in Different Firm Size Classes," Working Paper, Jönköping International Business School, 1996. Empirical evidence showing that equity-financed small firms are a key feature of the U.S. innovation ecosystem, serving as an effective mechanism for capitalizing on new ideas and bringing them to the market, was presented by Acs and Audretsch. See Zoltan J. Acs and David B. Audretsch, *Innovation and Small Firms,* Cambridge, MA: The MIT Press, 1990.

[20]For the first comprehensive review of SBIR, see National Research Council, *An Assessment of the Small Business Innovation Research Program,* Charles W. Wessner, ed., Washington, D.C.: The National Academies Press, forthcoming.

potential investors, SBIR awards improve the market for downstream investors. SBIR awards appear to have a "certification" function and by acting as a stamp of approval, help them obtain resources needed to grow from outside investors.[21]

Complementing the SBIR program, the Advanced Technology Program (ATP) was initiated in 1990 as a means of funding high-risk R&D with broad commercial and social benefits that would not be undertaken by a single company, either because the risk was too high or because a large enough share of the benefits of success would not accrue to the company for it to consider the investment worthwhile.[22]

At the conference, Marc Stanley, the director of the Advanced Technology Program, described his mission as one of bridging the gap between the research laboratory and the marketplace, emphasizing that ATP funding is directed to technical research but not product development. Companies, whether singly or jointly, conceive, propose, and execute all projects, often in collaboration with universities and federal laboratories. ATP shares the project costs for a limited time. Single-company awardees can receive up to $2 million for R&D activities for up to 3 years. Larger companies must contribute at least 60 percent of the total project cost. Joint ventures can receive funds for R&D activities for up to 5 years.

New Initiatives by State Governments

Participants at the conference also discussed new initiatives under way in the United States at the state and regional levels to bring home the benefits of innovation-led growth. Responding to the challenges of fostering regional growth and employment in an increasingly competitive global economy, leading American states have developed programs to grow companies as well as attract the talent and resources necessary to develop leading-edge technologies.[23] These state-based initiatives have a broad range of goals and increasingly include sig-

---

[21]With regard to the certification effect, see Joshua Lerner, "Public Venture Capital," in National Research Council, *The Small Business Innovation Program: Challenges and Opportunities,* Charles W. Wessner, ed., Washington, D.C.: National Academy Press, 1999.

[22]In a recent assessment of ATP, the National Academies noted that the program's cost-shared, industry-driven approach to funding promising new technological opportunities has shown considerable success in advancing technologies that can contribute to important social goals such as improved health diagnostics (e.g., breast cancer detection), developing tools to exploit the human genome (e.g., colon cancer protection), and improving the efficiency and competitiveness of U.S. manufacturing. See National Research Council, *The Advanced Technology Program: Assessing Outcomes,* Charles W. Wessner, ed., Washington, D.C.: National Academy Press, 2001, p. 87.

[23]For an overview of key state innovation initiatives, see The Pew Center on the States, *Innovation America: Investing in Innovation,* accessed at <*http://www.nga.org/Files/pdf/0707INNOVATIONINVEST.PDF*> on September 10, 2007. See also the publication by the Ewing Marion Kauffman Foundation and the Information Technology and Innovation Foundation, *The 2007 State New Economy Index.* Accessed at <*http://www.kauffman.org/items.cfm?itemID=766*> on September 10, 2007.

nificant resources, often in partnership with established companies and universities. As described by several conference participants, these efforts in support of regional development are playing an increasingly significant role in sustaining U.S. technological leadership.

One of the larger state efforts in this regard is the $300 million Texas Emerging Technologies Fund. According to Pike Powers of Fulbright and Jaworski, this fund is designed to help create jobs and to develop the economy of Texas over the long term by expediting the development and commercialization of new technologies and attracting and creating jobs in advanced technology fields. It focuses on increasing research collaboration through new Regional Centers of Innovation and Commercialization, matching research grants funds, and attracting more top-notch research talent to the State of Texas.

Another state-level initiative, introduced by Randall Goodall of SEMATECH, is the Texas Alliance for Nanoelectronics (TxAN). In his remarks, Dr. Goodall described TxAN as a statewide partnership for building innovative, virtual nanoelectronics capability "that leverages world-class researchers and R&D infrastructure and drives regional commercialization of technology." He described TxAN's new training paradigm, which includes a $4 million nanoelectronics development initiative to support 160 internships in advanced technology, a $3 million Nanoelectronics Research Initiative Center to provide university research in innovative ideas, and the Nanoelectronics Infrastructure Network that will link SEMATECH and TxAN with Texas universities in a $500 million collaborative effort.[24]

In addition to reinforcing policy successes at the state and federal levels, the United States can also learn from new policy initiatives under way around the world. As we see next, the innovation challenges facing Flanders and the policies adopted by the Flemish government to enhance competitiveness are directly relevant to the United States, just as U.S. experience may hold some implications for current policy development in Flanders.

INNOVATION IN FLANDERS

Flanders is a leading innovator of technology policy in Europe. Through a cohesive strategy that combines bottom-up input and top-down guidance with substantial public support, the Flemish government has promoted a technology-based national innovation system on a par with other highly effective global competitors such as Finland and Sweden.

---

[24]International SEMATECH R&D participation with regional governments includes the State of New York as well as the State of Texas. Expanding from its base in Austin, SEMATECH has put its new, $403 million research center in upstate New York. The state, in turn, has contributed in $210 million for equipment, construction, and specialized tools. New York has targeted nano as a key element in its future economic growth. In addition to $750 million in state funding in nano projects, many of them focused on the semiconductor industry, New York has received $7.25 billion in private investments. See Stephen Baker, "New York's Big Hopes for Nano," *Businessweek,* February 4, 2005.

Flanders' Innovation Advantages

This innovation strategy leverages Flanders' inherent advantages. As Peter Spyns of the Flanders Department of Economy, Science, and Innovation noted at the conference, Flanders' existing strengths include an excellent transportation and logistical infrastructure that takes advantage of Flanders' central location in Europe.

Flanders is also blessed with a well-educated, multilingual, hard-working populace. It benefits from a strong educational and research infrastructure that includes 7 universities, 22 non-university institutions of higher education, 5 university-based institutes of higher education designed specifically to diffuse knowledge out of the university, and several publicly supported research centers, including the Royal Academia and Musea. Flanders, he added, is also able to draw on substantial financial support from the European Union to develop its capacities in science and technology.

Flanders' Innovation Challenges

Flanders also faces challenges in innovation.

Technological Lags and Political Cycles. A major challenge in promulgating an effective innovation policy in Flanders—no less than for other democracies—is that the benefits of investing in innovation are usually realized over the longer term,[25] while popular expectations and the fortunes of elected policymakers run on shorter cycles. As Bart van Looy of the Faculty of Economics and Applied Economics at the Flemish Policy Research Centre for R&D Statistics (SOOS) explained at the conference, the time lag between investment and payoff poses a political liability.[26] Many people in Flanders expect the government to use its resources to create jobs directly and quickly, rather than to take the long and unfamiliar road of investing in R&D.

A Shortage of Seed Funding. Another challenge for Flanders is that even the most promising small and medium enterprises find it difficult to finance the developmental work needed to take a new idea to market.[27] Indeed, as Rudy

---

[25]Mark Myers of the University of Pennsylvania noted that at Xerox, where he had been director of research, the average time between the first expenditures on a new product to the first sales was 8 years; in pharmaceuticals, he said, this lag was about 13 years.

[26]In 2007, the Flanders government approved a second generation of policy research centers, resulting in, among other things, SOOS (Steunpunt O&O Statistieken) becoming SOOI (Steunpunt O&O Indicatoren).

[27]As Paul Ducheyne of the University of Pennsylvania noted in his conference presentation, the cost of initial research, particularly in the medical arena, is often dwarfed by the costs involved in developing and readying a product for the market. See a summary of his conference remarks in the Proceedings chapter of this volume.

Aernoudt, then Secretary-General of the Flemish Department of Economics, Science, and Innovation, noted in his conference presentation, venture capital firms, so familiar to inventors in the United States, are practically nonexistent in Flanders.[28] Because startups find so few private investors interested in taking a chance on a new business enterprise, he added, banks play an important role in providing private equity. However, as Professor Bruno de Vuyst of the Free University of Brussels (VUB) noted later, the amounts of capital the banks provide are also very small.

Cultural Aversion to Risk. Small firm formation in Flanders is also inhibited by a cultural aversion to risk and, more generally, relative inexperience with business formation and ownership. As a result, few individuals or groups are comfortable investing in new high-technology firms.

A Lack of Trust in Scientific Advance. A final challenge for Flanders, as for Europe, is a popular suspicion of some elements of science, such as genetic engineering, and especially genetically modified organisms (GMOs). "We have a population which is afraid of our industry," said chemist Erwin Annys of the Federation of the Belgium Chemical Industries in his conference presentation. "GMO is treated like a curse; nano-materials are frightening people. We need to work much harder on societal acceptance."

A STRATEGIC APPROACH TO PROMOTING INNOVATION

The Flanders government has sought to overcome these limitations with an integrated strategy (Box E) to prepare this small, but vibrant open economy for the rigors of global competition. The government's strategy is explained in a policy document, *Science, Technology, and Innovation,* which describes the roles that the Government of Flanders, the Federal Government of Belgium, and the European Union must play to support Flanders' innovation agenda.[29] The "great challenge," it says, is to turn Flanders "into a region where businesses establish their research centers and where high-tech companies can develop. The welfare and well-being of the Flemish people depend on this."

Providing Effective Political Leadership

The Flanders government is playing a leading role in developing policies to drive the future growth of its economy. Its leadership takes a sustained and

[28]Mr. Aernoudt resigned his post of Secretary-General of the Department of Economics, Science, and Innovation in September 2007.

[29]Under the Fifth Framework Programme, Flanders received more than 278.8 million Euros for research and innovation.

Box E
Flanders' Integrated Strategy for Innovation

Flanders is making significant investments in university training and in new government-funded structures to develop human resources, catalyze the commercialization of knowledge, and evaluate how well its various initiatives are working. Specifically, Flanders has designed a set of integrated strategies, including:

- Effective political leadership that is able to articulate challenges and is willing to provide resources commensurate with potential opportunities;
- Broad support for focused university-based research with incentives for patenting and commercialization;
- Programs that provide early-stage financing for SMEs;
- Systematic outreach, both in schools and via the commercial media, to explain the advantages of investing in research to drive the economy and raise the quality of life.

detailed interest in innovation policy. As Minister Moerman noted in her keynote address at the conference, "Investing in knowledge and innovation is crucial to economic growth." Mrs. Moerman offered a succinct description of Flanders' innovation strategy, which includes:

- Generous funding from the Flemish government and EU;
- Steady public encouragement and policy attention;
- Intermediary institutions, including public-private partnerships;
- State schemes to compensate for weaknesses in market mechanisms.

An important element of the innovation policy is an extensive publicity and public awareness campaign. The government takes advantage of newspapers, television, classroom lectures, school trips to research facilities, and other mechanisms to explain the importance of research, innovation, and the government's programs. Indeed, as Peter Spyns observed, "the word 'innovation' is everywhere these days. We are pushing its importance into the minds of people."

Encouraging Partnerships, Centers, and Networks

In recent years, Flanders has initiated a variety of new partnership programs described in a recent report authored by Greta Vervliet.[30] The Vervliet report

[30]Greta Vervliet, *Science, Technology, and Innovation,* Ministry of Flanders, Science and Innovation Administration, 2006.

describes multiple and overlapping mechanisms in Flanders to support both basic and applied research, create academia-industry-government partnerships, promote commercialization of new ideas, and inform the public about the values of research and innovation. Other stakeholders, including chambers of commerce and labor unions, are also assigned roles in this national scheme as well as their own innovation-enhancing organizations. These organizations include:

- Regional innovation "cooperation networks";
- Centers for collective research (these serve traditional industrial sectors);
- Competency "poles" (multidisciplinary centers, often located near universities);
- Four strategic research centers (for microelectronics, biotechnology, energy/environment, and broadband technology).[31]

Bridging Universities and Communities

An allied strategy has been to add to the traditional university functions of teaching and research, the task of bringing knowledge to the community. This function represents a radical departure from centuries-old academic custom. As Professor de Vuyst remarked at the conference, this third factor, "is a very big sea change," one that "brings new attitudes to these institutions."

Creating Non-Hierarchical Models of Collaboration

Emphasizing the importance of collaboration in advancing innovation, Mrs. Moerman described her ministry's effort to foster a "non-hierarchical" structure in which scientific and technological ideas flow from the bottom up. She also referred to the role of "horizontal administration," with both universities and institutes granted "a large degree of autonomy." Proposals for scientific work come from researchers, not policy makers, and are selected by traditional peer review. The government provides funding in the form of block grants to institutions that strive to meet performance metrics such as increased numbers of spin-offs, patent applications, and contracts. "Performance-based funding," she concluded, "is the key."

She also described a recent study by the Catholic University of Leuven, which sends, she said, "very positive signals" about the interaction between industry and academia. (See Box F.)

It is roughly estimated, she said, that in 2005 approximately 10 percent of all R&D expenditures in Flanders were generated by industries that were in collaborative partnerships with academia. This exceeded the figures estimated for the

---

[31]See the presentation by Peter Spyns in the Proceedings chapter of this volume.

Box F
A Win-Win Proposition for Knowledge Generation

A focus on commercial results does not necessarily reduce the amount or quality of basic research. Research conducted by Van Looy and Koenraad Debackere among others at SOOS has found that groups involved in tech transfer publish more, not less, basic scientific work.[a]

As Professor Debackere reported at the conference, "We found that groups that collaborate have a reinforcing effect and generate more fundamental scientific output as well as developmental research, as measured in number of publications. And industrial R&D feeds academia R&D in providing real problems."

---

[a]Bart Van Looy, Koenraad Debackere, et al., *Research Policy,* 2004. The researchers used data based on ISI-SCIE figures.

EU of 6.9 percent and the United States of 6.3 percent.[32] Also, she said, during the period from 1991 to 2004, universities and public research centers in Flanders had created 101 spin-off companies, 54 of them in the past 5 years.

Encouraging Innovation from Academia

Flanders has long encouraged the transfer of new ideas from universities to the marketplace. As long ago as 1972, the Flemish government allowed professors to reinvest their earnings from their inventions to create a more entrepreneurial climate in universities.[33]

In 1991, the Flemish Innovation Agency (IWT) was established as a "one-stop shop for innovation," offering direct financing for technology-related R&D and coordinating other innovation efforts of the Flemish government. IWT also provides services for new business and advice for the government.

In 2003, the Flemish government drew up an Innovation Pact between academia and industry, asking all parties to adopt the 3 percent Barcelona target.[34] The Flemish Science Policy Council (VRWB) was designated to monitor the Pact, using 11 key indicators. The first findings, published in 2005, found that Flanders was having limited success in transforming excellence in academic research activities

---

[32]European Commission, *3rd S&T Indicators Report,* Luxembourg: Office for Official Publications of the European Communities, 2003.

[33]See the presentation by Koenraad Debackere, Catholic University of Leuven, in the Proceedings chapter of this volume.

[34]This target challenges all EU nations to raise their total investment in R&D to 3 percent of GDP by 2010. According to the independent web portal EurActiv, however, this target is increasingly unlikely to be met (<*http://www.euractiv.com*>).

into useful and profitable applications. It also drew attention to the fact that a few international companies accounted for most of the region's industrial research.

Acting on these findings, Mrs. Moerman said, the government redoubled its efforts to spur innovation, beginning with a program to "reduce the innovation paradox" of good research and falling competitiveness. It was clear, she said, that the "old barrier" between academia and industry had persisted despite early bridging efforts. "There is still a strong feeling among academic researchers that working in industry corrupts a career and taints the principle of academic freedom," she told the conference. "And for its part, industry says that academia doesn't understand its needs."

Creating New Mechanisms for Technology Transfer

Accordingly, the government added two more innovation mechanisms. The first, in 2004, was an Industrial Research Fund (IOF) of €11 million to encourage universities to hire postdoctoral staff to perform research on findings with high potential for near-term market application; each university was allowed to decide on and create its own portfolio of industry-oriented projects.

The second mechanism places technology transfer offices (TTOs) at each university to help exploit research findings through spin-offs and patents and to provide advice to academic researchers on intellectual property issues.

The Flanders government efforts to bridge the university-industry gap do not stop there. At the time of the conference, the government additionally planned to enhance mobility among sectors by placing young academic researchers in industrial environments and support PhD students who plan to set up their own spin-off companies. It also planned to identify the scientific and technological areas with highest potential for future economic payoff, although Mrs. Moerman expressed some reservations about this process. She warned that the results should not be used to "reinstate thematic [top-down] priorities," which could be "an unfortunate return to the past when Flanders had several research programs defined from the top down. This didn't leave enough breathing space for bottom-up initiatives or for smaller research players."

Broadening the Concept of Innovation

In addition to addressing the innovation paradox, Mrs. Moerman noted that a second continuing policy challenge is to broaden the concept of innovation to include its non-technological aspects, including regulations, standards, training and education, patent and copyright issues, tax and economic policies, and labor market organizations. As noted in the Vervliet report, "innovation policy will be on the agenda of the Flemish Government as a whole."[35] Mrs. Moerman said that her

[35]Greta Vervliet, *Science, Technology, and Innovation,* op. cit., p. 19.

effort to "mainstream" innovation would include ensuring that it "becomes a horizontal dimension in all fields for which the Flemish government is responsible."

INVESTING IN WORLD-CLASS RESEARCH

The Flanders government provides direct support for *basic or investigator-driven research*, *strategic or policy-oriented research*, and *research with an economic focus*. As we see below, all three approaches are expected to play essential and complementary roles in Flanders' innovative strategy.

Funding Basic Research

The Vervliet policy document recognizes the importance as well as "unpredictable character" of basic research.[36] The Fund for Scientific Research-Flanders (FWO) provides funding for basic kinds of research through competitive peer review. Research topics are proposed by the investigators themselves, in a bottom-up strategy. In 2006, the core funding for the FWO consisted of €108 million, with another €11 million added from the National Lottery.

In addition, smaller amounts are allocated for international projects, United Nations University programs, international research facilities such as CERN, and the "Methusalem programme," which provides stabilizing long-term funding for senior researchers. Finally, in response to the general shortage of researchers in Flanders, the Odysseus Programme uses €12 million to attract top researchers from abroad, including expatriate Flemings.

Flanders' Strategic Research Initiatives

As Peter Spyns noted in his conference presentation, new firms in Flanders have lacked mechanisms for accessing knowledge developed in the universities. A large strategic research program, budgeted at €232 million for 2006, addresses this impasse by funding more basic research that generates knowledge for industry, the non-profit sector, and government and to strengthen research that is relevant to policy. The largest investments by this program support the three high-level research institutes:

- Interuniversity Microelectronics Centers (IMEC),
- The Flemish Interuniversity Institute for Biotechnology (VIB), and
- The Flemish Institute for Technological Research (VITO).

A new and fourth strategic research center, the Research Center for Broadband Technology (IBBT), was being formalized at the time of the conference.

[36]Greta Vervliet, *Science, Technology, and Innovation,* op. cit., p. 43.

These research centers represent innovative and, especially in the case of IMEC, world-class initiatives to encourage collaborative research.

IMEC

IMEC, established in Leuven in 1984, is the crown jewel of Flanders' research efforts. Recognized as a world-class microelectronics research center, IMEC strives to be a "worldwide center of excellence." As Anton de Proft, IMEC's Chairman, noted at the conference, it is "the world's largest industry commitment to semiconductor research in partnership—even though Belgians are hesitant to say they're the biggest anything."

IMEC emphasizes pre-competitive research and attempts to address the "innovation paradox" by bringing researchers from academia and industry together under the same roof. This provides focus for university researchers and basic solutions for industrial partners. Research subject areas include chip design, processing, packaging, microsystems, and nanotechnology. IMEC's stated mission is "to carry out R&D programs which are 3-10 years ahead of today's industrial needs."[37] In doing so, noted Mr. De Proft, IMEC consciously takes risks, but can afford to do so by sharing them among many partners. IMEC now has "core partnerships" with Texas Instruments, ST Microelectronics, Infineon, Micron, Samsung, Panasonic, Taiwan Semiconductor, and Intel, and "strategic partnerships" with major equipment suppliers.[38]

In July 2005 IMEC produced its first 300mm silicon disks with working transistors, using its second clean room, a new 3200-m^2 facility. A production ASML 170i immersion 193nm lithography system was installed in fall 2006, offering capabilities even beyond those available at the U.S.-based SEMATECH.[39]

Expressing the view of a U.S. core partner, Allen Bowling of Texas Instruments noted in his presentation that partnerships such as those with IMEC are now essential to sustain the semiconductor industry. He noted that a new product takes at least 4 years to develop fully, so that two or three products must be in development at one time, requiring more R&D capability than single companies have. One result is that costs of semiconductor research are increasing by more than 12 percent per year, while revenues are growing at 6.5 percent per year. Moving a new material or device into production requires 7 to 12 years of pre-competitive research, requiring the kind of intensive university input found at IMEC. "We leverage our dues substantially and gain great value from the IMEC focus on fundamentals," said Dr. Bowling. "There are more than 1,000 process steps in making an integrated circuit, so we need lots of help."

---

[37]IMEC Mission Statement.

[38]Greta Vervliet, *Science, Technology, and Innovation,* op. cit., p. 59.

[39]Allen Bowling of Texas Instruments, one of IMEC's international partners, personal communication.

Box G
Open Innovation and the IMEC Payback for Flanders

In a key exchange during the conference, Kenneth Flamm of the University of Texas asked Dr. de Proft what he termed an "impolite question" about the participation of large multinational semiconductor companies in IMEC. "Was not IMEC essentially subsidizing research for these firms, none of whom had production facilities in Flanders?"

Calling the question "astute and pertinent," Anton de Proft, the Chairman of IMEC, noted that the grants were not discounts on commercial contracts, but were meant to support fundamental research as a basis for further research programs with industry and with a view on long-term spill-over effects for the region.

When you dig deeper, he went on, you see payback for the region at many levels. He emphasized the presence of the residents, about 300 bright minds from around the world, spending their creative years here, and building up networks. They are all people likely to move up in their organizations, where they will be in positions to make decisions about where to put their R&D centers or other activities. Over 200 PhDs, he added, are performing their doctoral research at IMEC.

IMEC also is interacting with local industry and has created over 25 spin-off companies, among which are some very fast growers. IMEC's activities are also generating a strong secondary economic impact in the region, with over €42 million in subcontracting to the local industry.

The overall economic impact, he concluded, is a multiple of the government funding. "Our government is smart enough to understand not to look for direct matches, but to promote some formative behaviors without trying to steer the economy."

The 2005 budget of IMEC was about €235 million, about half of which came in the form of revenues from contracts with international industry; the remainder came from Flemish industry, the Government of Flanders, the European Commission, and several smaller organizations.[40] In all, IMEC has about 1,500 employees, including nearly 500 non-payroll industrial residents and guest researchers representing approximately 50 nationalities.

Flemish Interuniversity Institute for Biotechnology (VIB)

A second research institute that shows every sign of becoming another IMEC is Flanders' biotechnology facility. As Dr. Lieve Ongena, Senior Science Adviser to the Flemish Interuniversity Institute for Biotechnology (VIB), noted in her presentation to the National Academies delegation, the motivation for focusing on biotechnology is straightforward: "We had a lot of activity, but no transla-

[40]Greta Vervliet, *Science, Technology, and Innovation,* op. cit., p. 57.

tion from the universities to the economic growth of Flanders." VIB was given a compound mission designed to overcome that problem.

Formed in 1995 as a not-for-profit institute, VIB's mission is fourfold: to invest in basic research, to train young researchers, to commercialize discoveries, and to explain science to the public. It is an "institute without walls," staffed by scientists from Antwerp, Gent, Brussels, and Leuven. The plan has been successful, and the VIB now has 60 research groups in nine departments, and a 50/50 cost- and profit-sharing partnership with its four universities.

Addressing these missions, VIB has developed three core activities:

1. Biomolecular research focusing on molecular mechanisms of life. The broad objective is to concentrate on work of "strategic importance," including cancer research, cardiovascular biology, neurodegenerative disorders, inflammatory diseases, growth and development, proteomics, and bioinformatics.
2. An active patenting and licensing function whose goal is to transform the results of strategic basic research into industrial and social value.
3. A program to convey accurate and interesting information about science to the public.

The VIB supports 850 scientists and technicians, of whom 300 are PhD students. The total research budget is €60 million, half of which is a "strategic grant" from the government; the rest comes from the EU, the U.S. National Institutes of Health, and industry, with the proportion from industry growing.

The VIB uses two routes to transfer knowledge into societal benefits. For discrete discoveries, it may file a patent and license the technology to companies. If the "platform is wide enough," it may spin off its own company. The VIB has done this in four cases—for dVGen (using a microscopic worm for drug discovery), Peakadilly,[41] CropDesign, and Ablynx (using camel antibodies as a tool for drug targeting). Profits are used to promote growth and generate additional money for research.

A fifth startup, SoluCel, is a small company in Finland, and during the Leuven conference a sixth spin-off was announced, ActoGeniX, which uses a bacterium as a living drug delivery tool. To date, these startups employ more than 280 people and represent more than €220 million in venture capital.

To aid in firm formation, the VIB now plans to open its own small business incubator. Said Dr. Ongena, "If we have an idea, we want to be able to start a business tomorrow." To date, she added, the VIB had been efficient at commercializing, operating at the favorable cost of €1 million per record of invention and €2 million per patent.

[41]Since the time of the conference reported in this volume, Peakadilly (a biotech company located in Ghent, Belgium) has changed its name to Pronota.

Finally, the VIB places a high priority on communicating its work to society. The goal is to reach all levels of society, including the press and media, policy makers, teachers, students, doctors, patients, and scientists in other fields. For example, the "Scientists@work" school project invites groups of 10 to 15 students to the labs to work on a project for half a day. The immediate goal is to give them an authentic feel for careers in biology. The longer-term goals are to attract more bright students to science and to educate the public about controversial issues, such as the debate now raging in Europe over the safety of biotechnology.

Flemish Institute for Technological Research (VITO)

The third Strategic Research Initiative, VITO, is Belgium's largest and best-equipped multidisciplinary research center for energy, the environment, and materials. Its objective is to develop and encourage sustainable technological developments for government, industry, and SMEs. VITO seeks to do as much of this in partnership with industry as possible, and has recently increased contract research for industry to about 25 percent of income.

VITO has its roots in a Belgian agency started in 1988 to focus on nuclear and non-nuclear energy issues. It was overhauled in 1991 as an autonomous public research company, with the Flemish government as its sole shareholder. More than 80 percent of its work is performed on behalf of the Flanders Ministry of Environment and Energy. It has a staff of 510, 90 of whom hold PhDs, and a budget of €35 million for 2006.

As VITO's Managing Director, Dirk Fransaer, noted in his presentation to the National Academies delegation, VITO focuses on nine technology fields in addition to Exploratory Strategic Research (SBO) and strategic support tasks. The technology fields include decentralized energy systems, power technology, surface treatment, soil cleaning technology, innovative water purification, reactor technology, environment and health, air quality, and remote sensing. The SBO is medium-term research that aims to build up scientific capacity as a basis for economic and/or social applications.

Research Centre for Broadband Technology (IBBT)

This new center, opened in 2004, is a dispersed "virtual" center that focuses the missions of 13 existing research groups with the goal of becoming Flanders' fourth Strategic Research Initiative. It was founded on the premise that Flanders needs to be a leader in information and communications technology (ICT), and that to be a leader it requires large public investments in multidisciplinary basic research.

According to Dr. Waele, IBBT's general manager, the mission of IBBT is to develop multidisciplinary human capital and perform demand-driven research for industry and government. Its primary emphasis is on ICT innovations for the health care industry, which is seen to have the greatest potential for marketable

ICT uses. Research funded by the program will be primarily pre-competitive, requiring business partners to contribute at least 50 percent of costs. Major ICT firms working in Flanders include Philips, Siemens, and Alcatel.

IBBT intends to recoup its investments by both licensing and spin-offs. According to Dr. Waele, it does not seek to hold a portfolio of companies, but to create as many new firms as possible. Once a company has revenues, IBBT will take a low percentage—typically 5 percent.

IBBT gauges its success based on value added for companies and for the Flemish economy. It also uses secondary academic excellence indicators, as measured by spin-offs, and has plans to launch a business incubator.

IBBT has the freedom to work with foreign companies without restriction, as long as they are active in Flanders. "Our goal is to stimulate economic activity," said Dr. Waele, IBBT's general manager. "Borders are a thing of the past in terms of scientific collaboration."

Funding for Research with an Economic Focus

This last category of funding from the Flanders government goes to companies, research institutes or universities, and individuals who seem likely to promote "greater technological innovation in Flemish companies." They are designed to advance the goals of the Flanders government by:

- Creating conditions that increase technological innovation in companies;
- Creating conditions to achieve greater cooperation between academia and companies, and between companies themselves;
- Promoting a climate of innovation.

University Interface Services

The goal of this program, according to Mrs. Moerman, is to encourage universities to use their knowledge and expertise for the benefit of the Flemish economy and to develop a university culture where excellence in education and research is linked to innovative enterprises. The Flemish government supports university interface activities that encourage cooperation between university and industry and promote the creation of spin-off companies.

Cooperative Ventures

Funding for industry-initiated cooperative ventures includes various R&D subsidies for companies operating in Flanders and wishing to commercialize or otherwise add value to their research; support postdoctorate research; and create of economic networks that encourage innovation.

TABLE 1 Summary of Key Innovation Funding Sources and Institutes in Flanders

Innovation/Funding Agency	Role	Budget
Flanders Fund for Scientific Research (FWO)	FWO finances basic research, carried out in the universities of the Flemish Community and in affiliated research institutes.	Annual budget €119 million (2006) which includes approximately €50 million to fund individual researchers, €50 million to support Research Teams, and €2 million to promote Scientific Contacts.[a]
Flanders Institute for the Promotion of Innovation by Science and Technology (IWT)	Government agency providing funding for industrial and technological R&D, and technology transfer services.	Annual budget €240 million (2005) including approximately €75 million for R&D projects, €15 million for SME innovation projects, €37 million for strategic basic research, and €30 million for Cooperative Innovation Networks.[b]
Interuniversity Micro-Electronics Centers (IMEC)	The IMEC mission is "To perform R&D, ahead of industrial needs by 3-10 years, in microelectronics, nanotechnology, design methods and technologies for ICT systems."	The 2005 budget of IMEC was approximately €235 million, about half of which came in the form of revenues from contracts with international industry; the remainder came from Flemish industry, the Government of Flanders, the European Commission, and several smaller organizations.[c]
The Flemish Interuniversity Institute for Biotechnology (VIB)	VIB is a non-profit scientific research institute.	Total income of €62 million in 2006, with the Flanders government funding €31 million.[d]
The Flemish Institute for Technological Research (VITO)	Research organization to stimulate sustainable resource development.	€61 million in 2006. Own income generated is €32 million, with the balance of funding from government grants.[e]
Research Center for Broadband Technology (IBBT)	IBBT focuses on applied research in ICT in cooperation with companies and the government.	€17 million grant from the Flanders government.[f]

[a]Ministry of Education and Training, *Higher Education in Flanders, 2007*. Access at <*http://www.ond.vlaanderen.be/publicaties/eDocs/pdf/298.pdf*>.
[b]IWT Brochure. Access at <*http://www.iwt.be/downloads/publicaties/brochure/brochure_iwt_eng.pdf*>..
[c]Greta Vervliet, *Science, Technology, and Innovation*, op. cit., p. 57.
[d]VIB, *Annual Report 2006*. Access at <*http://www.vib.be/NR/rdonlyres/640AE8DE-5DCE-49F3-A46D-AD44E4AEAAE1/0/VIB_AnnualReport2006b.pdf*>.
[e]VITO, *Annual Report 2006*. Access at <*http://www.vito.be/english/who/vito_en2006.pdf*>.
[f]IBBT, *Annual Report 2006*.

Support is also provided for developing sustainable technologies, preparing spin-off companies, and incentives to encourage innovation in SMEs. The Flemish Innovation Cooperative Ventures (VIS) program supports collective research, technological services, projects that simulate innovation for particular issues, and activities to stimulate subregional innovation. For 2006, the budget of the program was estimated at €160 million.

New Experiments in Financing R&D

In 2001, the government created a program called *Arkimedes*, which tries to overcome cultural aversions to risk by providing government guarantees and tax credits for people who invest in several kinds of small-denomination bonds. As described by Rudy Aernoudt, the money raised by these bonds ("a pool of pools") goes into several R&D funds, whose effectiveness is measured by the number of innovative companies produced. Risk is said to be low because the loans are spread among numerous companies, although the program is still too young to draw firm conclusions about its effectiveness.

THE ROLE OF THE CATHOLIC UNIVERSITY OF LEUVEN

The Catholic University of Leuven (K.U.Leuven), the oldest university in Belgium, plays a significant role in Flanders innovation strategy. K.U.Leuven's R&D mission is "to promote and support knowledge and technology transfer to industry."

According to Professor Koenraad Debackere of K.U.Leuven, this mission is carried out at three levels. At the top are researchers on the payroll. As of 2005, he reported, K.U.Leuven supported 974 researchers, a number that had doubled in 5 years. Many of these do research for industry. At the middle level, the university is actively involved in three areas: contract research, spin-off formation and regional development, and IPR and licensing. The third level is industry itself.

Traditionally, he said, the university had two basic missions: research and teaching, which are still fundamental. But in that traditional academic environment, faculty research was done in "almost a pure ivory-tower setting." Nowadays, however, universities in many European countries are charged by the government to create structures and activities that support the commercialization of their research.

K.U.Leuven's Matrix Structure

At K.U.Leuven, which Dr. Debackere considered an unusual case for Europe, there is a "full matrix-like structure" that gives academic researchers incentives to collaborate with industry. The academic subjects are divided into three groups: biomedical research, the other exact sciences, and the arts and humanities. Within each are the faculty members and the different departments, "the

Box H
Growing the University's Economic Role

According to Dr. Debackere, Leuven's success at commercializing R&D is based on:

- A critical mass of high-quality, internationally competitive research. "This is why IMEC is very strict in its performance assessments."
- An integrated approach to technology transfer, such as incentives for multi-disciplinary teams and high value-added services.
- Clear incentives and policies to encourage individuals, research groups, and departments to pursue spin-off opportunities.
- Creation and acceptance of an entrepreneurial climate in a university context.
- A Flemish legal context that is positive with respect to the exploitation of academic research and IP.

normal hierarchy where people are recruited and promoted on the basis of their teaching and research abilities."

At the same time, the university has a horizontal structure with about 50 research divisions under the umbrella of a central office of R&D. The divisions are organized on an interdepartmental basis, and professors of research become members of one of those divisions, under which they can organize their industrial involvement. Any proceeds from their work remain within the division. What drives them, said Dr. Debackere, is a desire to be part of a strong research environment where they can compete and collaborate with the best of their colleagues.

In order to promote a strong collaborative research environment, the university lets the faculty reinvest the income in infrastructure, equipment, and postdoctoral scholars. "Although this has been criticized as 'social welfare'," he said, "we regard it as the best kind of social welfare, because everything is reinvested in the research." In order to support the divisions and their activities—which include applied research, technology transfer, and the generation of new companies—about 40 people are employed to provide management support, IT support, and consulting on the incubation of new companies.

"Leuven Inc."

Dr. Debackere said that in Leuven, more than 100 spin-off companies had already been created, leading to the nickname "Leuven Inc." Part of its success in expanding entrepreneurship, he said, grew out of the formation of effective networks. Some of these were horizontal: contact between universities, IMEC, startups, and other "innovation actors." Others are vertical: technology clusters, such as DSP Valley that focus on design of hardware and software technology for digital signal processing, and L-SEC (Leuven Security Excellence Consortium), an international non-profit network dedicated to promoting the use of e-security.

EVALUATING THE IMPACT OF POLICY INSTRUMENTS

Given the challenges in accelerating innovation in Flanders, and in the European Union more generally, Flanders pays special attention to evaluating its efforts to spur innovation. It has found that despite the magnitude of its investment, not all efforts are fully successful.

The Flanders government in 2000 charged SOOS, the Policy Research Center for R&D Statistics, with answering such questions as whether the new commercializing role assigned to universities would add value for society and whether it would crowd out private investment.

So far, the Policy Research Center has found a positive impact in patenting activity and increased technological activity and no crowding out effect, as tested by numbers of transfers of ownership rights. In all, reported Professor Van Looy, "the findings suggest a distinctive and considerable positive impact."[42] He added that more important than any single mechanism would be the sustained long-term political commitment of the government.

Another evaluation study attempted to identify factors that produced successful new firms, and found some ambiguous answers. They found, for example, no "straight relationship" between equity financing and growth. They also found that rapid growth correlated with high failure rates. Small firms benefited from having teams of two or three founders, whose members had commercial experience, but more important seemed to be early involvement in international activities. One researcher observed that most Flemish policy measures have been designed to address the equity gap, but the mix of human resources is overlooked, and early-stage internationalization is the key.[43]

CONCLUSION

The United States and Flanders differ enormously in scale, politics, and culture. The U.S. population is about 50 times that of Flanders, and 30 times that of all Belgium. The people of Flanders assign a more prominent role to government, take a cautious view of risk-taking, and experience relatively little venture capital activity. Even so, the Flemish government has found that the process of innova-

[42]See presentation by Bart Van Looy in the Proceedings chapter of this volume.

[43]B. Clarysse, Policy Research Centre of Entrepreneurship, Enterprises and Innovation, conference presentation.

Box I
Growing a Regional Innovation Economy

According to Luc Soete of the University of Maastricht, four conditions are necessary for stronger innovation-led growth and development. Most of these, he said, are already in place in Flanders:

1. High-quality human capital formation. Core elements for Flanders, he said, were universities, polytechnics, and professional training schools, including lifelong learning programs. These emphasized high quality, reduced failure and dropout rates, improving attractiveness to students from other regions and use of exchange programs as benchmark learning tools.
2. Open research practices. "IMEC is the clearest example of this," he said. "Texas Instruments brings eight people here, and assumes that they learn as much as they 'leak'. This openness attracts people." It also strengthens the research presence, stimulates joint public-private initiatives, benefits from "foreign" knowledge and collaboration, and strengthens the regional research infrastructure.
3. Stronger innovation performance. He emphasized the importance of supporting local science spin-offs and entrepreneurs, for which Flanders has created specific policies. Flanders has also strengthened innovation by linking public research institutions, teachers, and local SMEs; embedding large multinational corporations in the public research infrastructure; and sponsoring public information projects to explain innovation.
4. Regional capacity to absorb innovation. Flanders' support for regional "beta users," or early adopters helps grow the seeds of innovation. Capacity absorption is also hastened by procurement policies, a regional presence abroad (e.g., at fairs), a focus on regional diffusion of knowledge, and cooperation with other "foreign" regions.

tion seems to be less a function of scale than of human resources, a conducive environment, and political will.

While many of the Flanders region's policies and programs to support innovation are too recent to allow conclusive evaluation, some outcomes are already apparent. These include an increase in numbers of spin-off companies, high numbers of publication and patents in biotechnology, and the growing reputation and impact of microelectronics research conducted at IMEC. These initiatives, described in the conference proceedings found in the next chapter, are worthy of broader notice.

Some of the policy measures discussed at the symposium may be of interest to countries and regions around the world, although this would normally be for adaption and adoption, rather than direct copying. To adapt them to the specific contexts and conditions of different national or regional innovation systems, it is necessary to understand considerably more about the specific designs of the different policy measures discussed, as well as their roles in their specific innovation system and policy contexts. In addition, innovation policies and programs that address important challenges must be scaled in relation to the entire system or parts of the system they address. Innovation policies and the resources devoted to them often suffer from a "tyranny of small scale." Even well-conceived programs cannot make a meaningful contribution to innovation performance unless the program and resources allocated are adequate to the task.

Taking into consideration these caveats, policymakers in the United States can find instructive lessons in the broad goals, multiple instruments, significant funding, sustained activity, and regional branding found in the Flanders experience. Such a comparative perspective is essential if we are to respond manfully to this century's innovation imperative.

II

PROCEEDINGS

Welcome

Peter Spyns
Department of Economy, Science, and Innovation
The Flemish Government

Dr. Spyns welcomed Flemish and U.S. attendees to the symposium with an introduction to his department and to the Flemish government's innovation system. He said that the objective of his department was to build up strategic intelligence in the area of innovation policy preparation. One approach to this task is to exchange information with government agencies abroad, he said, and he described the current international conference as "a very good way of realizing this information exchange."

He noted that innovation theorists today speak in terms of an innovation system, in which knowledge is distributed among all participants. This was another good reason for organizing a bilateral conference, because it facilitated the sharing and disseminating of knowledge. Because the invitees included people from academia, government, industry, agencies, and advisory bodies, he said that the effective sharing of knowledge was already assured.

He gratefully acknowledged the conference host, the Interuniversity Micro-Electronics Centre (IMEC) in Leuven, which he described as "one of our best world-wide research centers. The fact that we have this conference here underlines their success in achieving international excellence, and therefore this location is very appropriate for this conference."

He gave a brief overview of the symposium sessions to come, and invited Dr. Spencer to offer introductory remarks on behalf of the U.S. delegation.

Remarks on Behalf of the U.S. Delegation

William J. Spencer
SEMATECH (retired)

Dr. Spencer introduced the Flemish audience to the Science, Technology, and Economic Policy (STEP) program, the co-host of the conference, as part of the U.S. National Research Council, the "working arm of the National Academies." This symposium was part of a series of workshops that had begun about 18 months earlier and included meetings in Taiwan, China, and India. "Europe is and will continue to be a major economic player," he said, "so learning from what you've done is important."

He recalled that he had first come to Europe about 41 years earlier on a journey that included a drive to a conference in Liège. The entire town was dark because it had no street lights. He and his colleagues had been able to find the house by a sliver of light that escaped under the front door. He remarked on how much had changed since that time. On the flight over the Atlantic he had thought about the new frontiers of technology that had opened, beginning with the invention of the transistor in 1947 and the integrated circuit. When he first visited, the only uses of the IC were a few defense applications. All arrangements for the meeting in Liège had been made by telephone or snail mail; the current meeting, by contrast, had been arranged entirely by email, thanks to the wide new communications universe of the Internet.

Since 1965, he said, Europe had taken giant strides, and was poised to take more. He recommended the book by T. R. Reid, *The United States of Europe: The New Superpower and the End of American Supremacy*.[1] "In 1965 we came to lecture. Today we come to listen and learn."

---

[1]T. R. Reid, *The United States of Europe: The New Superpower and the End of American Supremacy*, Penguin Press, 2004.

Session I

Perspectives on the Flemish Innovation System

Moderator:
Charles W. Wessner
U.S. National Research Council

THE FLEMISH INNOVATION SYSTEM AND ITS COMPONENTS

Peter Spyns
Department of Economy, Science, and Innovation
The Flemish Government

Dr. Spyns began with a description of Belgium's unique constitutional arrangements, which provide Flanders with significant autonomy in setting policies to encourage innovation. He outlined the policymaking apparatus that includes the Flemish Parliament, the Ministry for Economy, Enterprise, Science, Innovation, and Foreign Trade, and the Department of Economy, Science, and Innovation within the ministry. Advising the Flemish Parliament and the government is the Strategic Advisory Council for Economy, Science, and Innovation (VRWI). Moving beyond the policymaking apparatus, Dr. Spyns listed Flanders science and technology agencies, including the Fund for Scientific Research (FWO), a grant giving agency, and the Institute for the Promotion of Innovation through Science and Technology (IWT), which he described as a "one-stop shop for technological research and innovation support in Flanders." These and other economic agencies, advisory councils, and technology assessment institutes provide a framework for a complex innovation system.

Flanders has a vibrant industrial sector. An open economy situated in the heart of western Europe, Flanders' key industries include chemicals and pharmaceuticals, auto manufacturing, food and beverage processing, and the diamond trade (among others). Stakeholder organizations include the Flemish network of Chambers of Commerce and Industry (VOKA) and Agoria, Belgium's largest employees organization and trade association for the technology industry.

Turning next to the education and research system, Dr. Sypns, listed Flanders' 7 major universities in addition to 22 institutes of non-university higher education and other associations, private colleges, and research centers. Also active in this arena are the Flemish Interuniversity Council (VLIR), which is an umbrella consultation body between the Flemish universities and the Belgian authorities responsible for higher education and research, and the Flemish Council for Non-University Higher Education Institutions (VLHORA).

Bridging innovation at universities and industry are several intermediary organizations. This includes *strategic research centers* like IMEC for semiconductor and nanotechnology, VIB for biotechnology, and VITO for environmental technologies; 11 *competence poles* to bring multidisciplinary focus to research in technologies related to food, logistics, materials, and cars; 15 *centers for collective research* that addresses the needs of traditional industries; and 5 *university interface groups*.

In all, while Belgium's federal structure provides strong regional autonomy to shape Flanders's innovation system, sustained, high level attention to innovation policy by Flemish policymakers has been equally (if not more) important. The result, Dr. Spyns concluded, is a "relatively well-performing Flemish Innovation System."

IMPLEMENTING AND MONITORING THE FLEMISH INNOVATION SYSTEM

Eric Sleeckx
Flanders Institute for the Promotion of Innovation by Science and Technology (IWT)

Mr. Sleeckx discussed evidence that money spent by the Flanders government on innovation is providing a good return on investment. He began with some key figures, including the annual budget provided by the government of €250 million, about €90 million of which goes to R&D projects at subsidy levels of 25-50 percent. An additional €15 million goes to about 400 innovation projects of small-to-medium-sized enterprises (SMEs), €37 million to "strategic basic research," €7 million to higher education research, and €30 million to cooperative innovation networks. Altogether, he said, IWT works with about 150 large enterprises per year and 500 SMEs.

He said that the rationale for public support of innovation was that government can and should compensate for inherent market failures that hold economies back. "Subsidies alone provide a kind of leveling," he said. "But there is more and more pressure to prove that subsidies are active drivers of innovation, so we have a firm focus on additionality."

Positive Results from an IWT Study

He summarized the results of an IWT study that supports this position.[2] The study found that:

- Firms invest 100 percent more on R&D when subsidies are available.
- Firms considered innovative spend 53 percent more than non-innovative firms.
- For IWT funding, the effect of €1 of additional funding leads to €0.85 to €1.34 in added R&D spending at the firm level.
- The injection of public funds does not crowd out private investment, and the full amount of subsidies was spent for R&D.
- The impact of the program was clearer for SMEs than for larger firms.

He also posed the question, "Can government intervention cause firms to change the way they do R&D in a desirable direction?" Again, the answers, especially for SMEs, had been generally positive, including a series of preliminary results:

- Forty percent of R&D projects would not have occurred without subsidies.
- Seventy percent of firms undertake regular R&D&I projects after receiving IWT subsidies.
- Firms that received subsidies tend to return to the program for additional subsidies.
- Subsidized projects are more ambitious and of larger scale.
- The program had positive effects for SMEs in involving external knowledge centers.
- Firms may gain limited "competence additionality" (innovation skills), but these spilled over to non-subsidized projects.
- Subsidies enabled firms to undertake desired R&D sooner.
- Thirty percent of product innovations and 38 percent of process innovations would not have occurred without IWT subsidies.

---

[2]Kris Aerts and Dirk Czarnitski, "The Impact of Public R&D Funding in Flanders," *IWT Studies,* 54. Access at <*http://www.iwt.be/downloads/publications/observatorium/obs54.pdf*>.

He said that the IWT studies indicate that "it makes sense to subsidize R&D. In fact, the results seemed almost too good, so that additional checking will be done." In general, however, he said that IWT had found that subsidies have a "clear and convincing positive impact on SMEs." The evidence for large firms, he said, "was not as convincing."

He then described the Cooperative Innovation Networks (CINs), by which Flanders distributes its innovation support. These networks share the following characteristics:

- Ideas are generated from the bottom up: Firms propose their own projects and request the support.
- IWT was then subsidizing about 110 projects—typically of 4-year duration, with the option of extension.
- Some 250 innovation advisers are employed to offer support to these firms, at a cost of about €25 million per year.
- About 85 different organizations take part in the networks, including federations, R&D centers, employers' organizations, and "company clusters."

Evaluating a New System

Again, Mr. Sleeckx reported considerable effort on evaluation. The evaluation system ("RAP") reported on what each network did, and assembled web-based reporting of activities to judge how well each project was performing. Each project defined its target values for a subset of RAP numbers, and all activities were reported three times a year; e.g., a CIN reports 15 seminars, 45 company visits, etc. CINs are encouraged to include up to four success stories in their report.

"The system has been running for about 3 years," said Mr. Sleeckx. "It is working quite well and is accepted by project leaders." The evaluation system, he said, reduced reporting efforts, shortened follow-up time for IWT, and allowed easy identification of problems.

In a next step of optimizing the reporting for the CINs, IWT asked such questions as whether they are hitting their target, what are the effects, and does the company use outside advice. It was not intended as a measurement of economic benefits for companies, but a way of judging whether the CIN is providing the right services. He concluded by listing seven best practices that have been identified in building up this follow up system:

- Developing the right tools and standardizing them are important for monitoring.
- The tools must be developed in cooperation with the firm's players or will not be used.
- The focus must be on benefits for participants.

- Communication and behavior by the administrators must be consistent.
- Changes must be introduced step by step, since they require changes in culture.
- IWT must give immediate feedback to companies to prove that the data are used.
- IWT must add value to the data and return them to the reporting organizations.

Discussion

Dr. Wessner asked whether the R&D subsidies were given as salaries, tax rebates, or other forms. Mr. Sleeckx said in Flanders both of those forms are used; IWT provides only direct grants.

A questioner asked what was expected from each company. The answer was that "they have to prove value for money. The rule is that they must generate at least ten times the money they get from us, or 25 times when abroad." A participant asked how an American company would be expected to valorize money outside of Flanders. Mr. Sleeckx said that IWT subsidizes U.S. companies only within Flanders.

Dr. Myers asked for more detail about specific economic failures and whether they had been corrected. Mr. Sleeckx answered that the results are hard to measure, and would be more apparent in 5-10 years.

Dr. Spencer asked about total R&D spending in Belgium. This is difficult to calculate, answered Mr. Sleeckx, although the country, like the rest of Europe, is targeting 3 percent of GDP for R&D. Some 2 percent would be spent by government and 1 percent by industry. The amount now is thought to be about 2.4 percent, at the same ratio, with some federal money included.

CURRENT EU INNOVATION POLICY CHALLENGES: FROM LISBON TO LOUVAIN

Luc Soete
University of Maastricht, Netherlands
& UN Univ-MERIT

Professor Soete said that he would try to give an EU policy summary despite never having been an EU official. He reviewed some general features of EU innovation policy, saying that the EU framework for 2006-2007 had been launched with high expectations, including a new coordination mechanism called Open Method Coordination (OMC) for areas outside the European treaty. He said that there are areas of easy reform, and areas that are more difficult. The knowledge area was one of the easy areas, including R&D and innovation. At the EU level, on the other hand, areas of reform were moving at a rate of "slow to no," especially in areas such as constitutional reform, budget, institutions, service directives, agriculture, and patents.

The EU was experiencing the same competitive pressures felt by the United States, he said, notably global competition from the BRIC (Brazil, Russian, India, China) countries, which was increasing rapidly. Troublesome policy issues grew out of the activities of some competing countries that favored growth in ways that were non-democratic or environmentally unsustainable. China, in particular, had an enormous appetite for natural resources, with negative terms of trade for the EU. The EU, in turn, had the disadvantages of an aging workforce and a shortage of knowledge workers.

The Need for Structural Reforms in Europe

Meanwhile, Europe had strong needs for structural reforms, including a Lisbon agenda to deliver welfare and employment. A large percentage of EU citizens were still in need of meaningful work, and economies needed more activities that added economic value. The EU's productivity gains were lagging, as the industrial and services structure, encumbered by too many rules and regulations, had difficulty competing with emerging nations. The current rate of development in Europe seemed at present unsustainable, not just from an environmental perspective, but also because of the demands for social care and health care for an aging population.

Internally, growth was unsatisfactory. Lisbon 2006 showed the striking lack of internal growth dynamics in the EU—a combination of apparently sound macro-economic policies but few incentives for structural reforms. Without growth-enhancing policies, he said, science and engineering knowledge capital was growing too slowly, leading to the emigration of scientists and engineers, lagging public investments in knowledge, and outsourcing of private knowledge activities. He cited the U.S. economist Richard Freeman, who had written that the EU, with 70,000 PhDs, should have a 1 percent higher growth rate than the United States, which has 40,000 PhDs.

A Lag in Growth Rate

Instead, the growth rate was lagging the United States by several measures. He said that labor productivity in particular had become a major issue. Labor productivity for the EU-15, as a percentage of U.S. labor productivity expressed in GDP per hour, had been rising for several decades, but this trend had slowed through the 1990s. In 1998 the EU had the same labor productivity as the United States. Since then it had declined in Europe and risen in the United States.[3]

[3]Trend growth of annual growth in GDP per hour worked, U.S. and EU-15, 1979-2004, van Ark, with Hodrick and Prescotte, 1987.

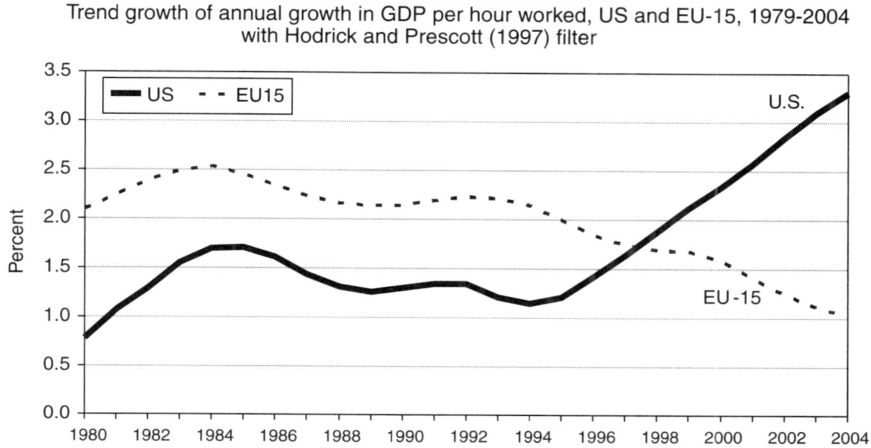

FIGURE 1 The EU productivity growth slowdown.

Historically, he said, Europe had reached the end of catching up, and GDP per hour was now declining. In addition, Europe suffered a major per capita income gap in relation to the United States. One factor causing this gap, he said, was the slowdown in European private R&D expenditures, "which appears structural." Industry in the United States spends twice as much on R&D as the EU25 member countries, and had increased until late 1990s. That gap had narrowed somewhat, but was still there. "It is clear," he said, "and it is a structural gap."

He then criticized the "R&D obsession" of Europe, exemplified by the Barcelona target of 3 percent of GDP. This target, he said, is "too narrow, too soft, and too passive." He did agree, however, that the core problem for the EU was one of private investment in knowledge, especially tacit knowledge. Reliance on public funds, he said, was justified in the continental and northern EU countries from the perspective of equal access and was consistent with progressive income taxation; higher education, for example, was virtually free. However, over the 1990s, the tax burden had been significantly reduced for both businesses and high-income citizens, so that almost no one would pay a rate of more than 50 percent for all taxes.

In fact, he said, "free" access to public knowledge resulted in increasing inequality and crowding out of private knowledge investment, which had no incentive to invest in the universities as long as they were so generously supported by public money. And even this degree of public support was insufficient. For while Europe has almost exactly the same number of universities and polytechnics as the United States, every university in Europe is underfunded by approximately half in comparison with those in the United States. "The primary need," he said, "is to find incentives for the private sector and individuals to invest back into the universities."

Europe as a "Unique Social Laboratory"

Professor Soete moved to a more general discussion of regional policies in Europe and the emerging importance of knowledge, which he said was essential at a time of decreasing social cohesion. "In fact," he said, "there is so much variation in the regions one could argue that knowledge is a replacement for social cohesion." He called Europe a "unique social laboratory," and urged greater speed in the ongoing shift from traditional industry to R&D and innovation policies with a strong impact on growth, structural change, and international competitiveness. This was happening far later than foreseen by many economists half a century ago, who predicted that a backward post-war Europe was likely to catch up in the 1950s.[4] Even today, he said, Europe is held back by its complex web of customs, rules, and social expectations. The recent Aho Report[5] was especially sharp in questioning European policies and structures, including the following criticisms:

- Support of industrial/R&D/innovation policies is insufficient.
- Support for those policies is further weakened by insufficient industrial renewal and industrial R&D investment.
- The EU policy of funding projects until the country reaches its preordained level is arbitrary and detrimental (work on a bridge stops even if it is not finished).
- Europe lacks large industrial complexes.
- Standards and national regulations are fragmented.

From a regional perspective, he said, four situations are needed for stronger growth and development, most of which are already in place in Flanders:

---

[4]The period 1950-1973 has been described as a "golden age" of Europe, including extraordinarily rapid economic growth based on heavy investment and enhanced social capability for growth. It has recently been suggested that more research is needed on the quality of human capital, openness of research capacity, and the role of institutions in influencing rates of return on investments. E.g., N. F. R. Crafts, "The Golden Age of Economic Growth in Western Europe, 1950-1973," *Economic History Review,* 3:429-447, 1995.

[5]The Aho Commission, chaired by former Prime Minister of Finland Esko Aho, urged Europe's leaders to take radical action on research and innovation "before it is too late." In a report released January 20, 2006, it proposed a four-pronged strategy focusing on (1) the creation of innovation-friendly markets, (2) strengthening R&D resources, (3) increasing structural mobility, and (4) fostering a culture which celebrates innovation. Access at <*http://ec.europa.eu/invest-in-research/action/2006_ahogroup_en.htm*>.

1. High-quality human capital formation. Core elements for Flanders, he said, were universities, polytechnics, and professional training schools, including lifelong learning programs. These emphasized high quality, reduced failure and dropout rates, improving attractiveness to students from other regions, and use of exchange programs as benchmark learning tools.

2. Open research practices. "IMEC is the clearest example of this," he said. "Texas Instruments brings eight people here, and assumes that they learn as much as they 'leak'. This openness attracts people." It also strengthens the research presence, stimulates joint public-private initiatives, benefits from "foreign" knowledge and collaboration, and strengthens the regional research infrastructure.

3. Stronger innovation performance. He emphasized the importance of supporting local science spin-offs and entrepreneurs, for which Flanders has created specific policies. Flanders also strengthened innovation by linking public research institutions, teachers, and local SMEs; embedding large multinational corporations in the public research infrastructure; and sponsoring public information projects to explain innovation.

4. Regional capacity to absorb innovation. He emphasized Flanders' support for regional "beta users," or early adopters, in helping grow the seeds of innovation. Capacity absorption is also hastened by procurement policies, a regional presence abroad (e.g., at fairs), a focus on regional diffusion of knowledge, and cooperation with other "foreign" regions.

Professor Soete concluded that regional innovation support policies, such as those of Flanders, will become ever more critical for Europe, especially to catalyze social cohesion among diverse countries. He called for more EU-sponsored fundamental and strategic research for all 25 member countries and a larger role for universities and research institutes in generating and applying technology. Through interaction and collaboration, he said, these "hotspots" will learn from each other and raise underutilized growth potential across national borders. This movement was likely to spread around the world, he said, as the notion of national competitiveness becomes outdated and gives way to "a world-wide explosion of technological hotspots."

Perspectives on the U.S. Innovation System

Moderator:
Luc Soete
University of Maastricht, Netherlands
and UN Univ-MERIT

CHALLENGES AND CURRENT DEVELOPMENTS IN THE U.S. INNOVATION SYSTEM

Mary Good
University of Arkansas at Little Rock

Dr. Good said that because she was a scientist, her talk would focus on micro-economic aspects of the innovation system in ways that might complement the macro-economic views of Professor Soete, who was an economist.

She began with the following definition of innovation:

> "Innovation is a strategy that provides resources to talented people in an atmosphere which promotes creativity and is focused on outcomes ranging from new products, to customer satisfaction, to new scientific insights, to improved processes, to improved social programs. [It is] designed to create wealth and/or improve the human condition."

Why, she asked, does innovation matter so much in a global economy? In the United States and Western Europe, she said, the standard of living had been built on innovational competition. In particular, the U.S. position in a "free market" has depended on productivity: the ability to take risks, especially in new enterprises, and an instilled belief in upward mobility. It has provided higher wages for those who "work smarter" and allowed the creation of new wealth for risk takers.

Elements of the Innovation System

In the current and future global economy, she said, many new competitors are emerging, thanks to low wages, a focus on education in science and engineering, and creative ways to attract capital. Countries now know that innovation requires an interlocking set of priorities, which she listed under the following outline.

Talent. Each nation needs a strong educational system and a motivated workforce with diverse skills and interests, as well as a dedication to lifelong learning. Emerging technological powers were creating cadres of technical professionals "capable of inventing the next game-changing technological wave and exploiting the current knowledge base, wherever it exists."

Investment. Each society must provide resources for long-term development of new, unexplored areas and for short-term development of improved products, processes, and services.

Infrastructure. Physical environments are needed that are conducive to state-of-the-art exploration and business conditions that encourage risk-taking and collaborative activities. These include IP protections, health care, and energy certainties.[6]

Innovative societies also need a culture that values and rewards risk taking and tolerates failure, she said. The venture capital community often favors people who have failed, in fact, because they assume that failure is an effective teacher, equipping them to meet the next challenges.

Challenges for Innovation in the United States

The United States faced several issues in optimizing its innovation capacity, she said, especially that of demographics. The population is adding new young people, many of whom are minorities with little education. Also, the country faces a complex challenge in admitting educated newcomers while restricting illegal immigrants or those who wish the nation harm. Finally, as many of the current generation of scientists and engineers begin to retire, the nation must learn to accommodate their longer life spans and transfer their knowledge to the next generation.

---

[6]Adapted from Council on Competitiveness report, *Innovate America,* Washington, D.C.: Council on Competitiveness, 2005.

Another difficult issue is education, especially K-12 public education. The level of math and science skills varies widely by geographic and economic status, and the skill of educators. For the public community colleges, which must educate many first-generation Americans, challenges exist, including teaching at a level adequate to allow students to move to 4-year science, math, and engineering curricula. For the public universities, state funding has declined and the schools have difficulty financing lower-income students.

She noted an issue that overlaps with education, which is the presence of large numbers of foreign-born students in many fields of science and engineering. In 1994, the U.S.-born graduate population in scientific and engineering departments was far higher than the foreign-born population, but that has dramatically changed. Now about half of S&T graduate students are foreign-born; in engineering, 65-70 percent are foreign-born.

U.S. innovation depends on the availability and continued presence of these foreign-born students. But will they stay, she asked, as other countries quickly build up their own research universities and job opportunities and our own immigration system discourages them from staying?

Another factor weighing on the U.S. innovation system is declining investment in R&D by all sources. While the federal investment has risen in constant dollars since 1976, almost all of this increase has gone to the defense sector. Spending on non-defense research rose in the 1990s, with most of the increase

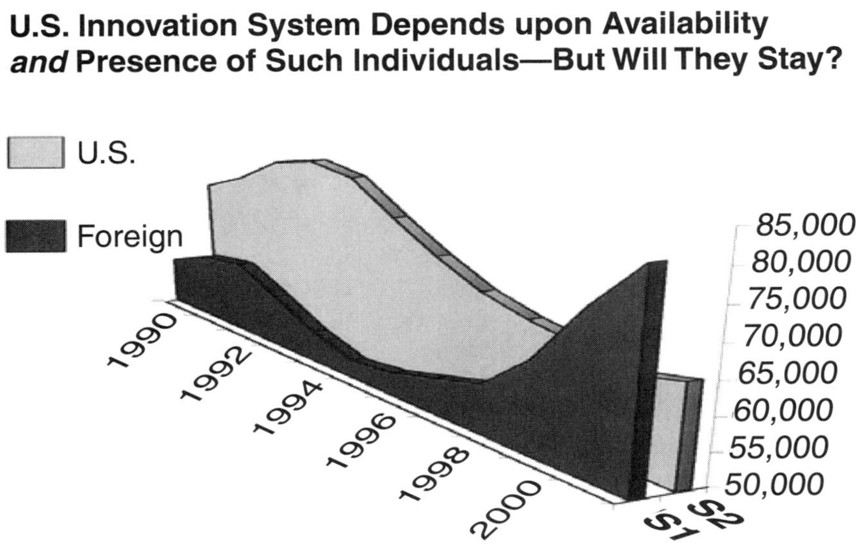

FIGURE 2 Foreign-born students awarded majority of U.S. scientific graduate and PhD degrees.

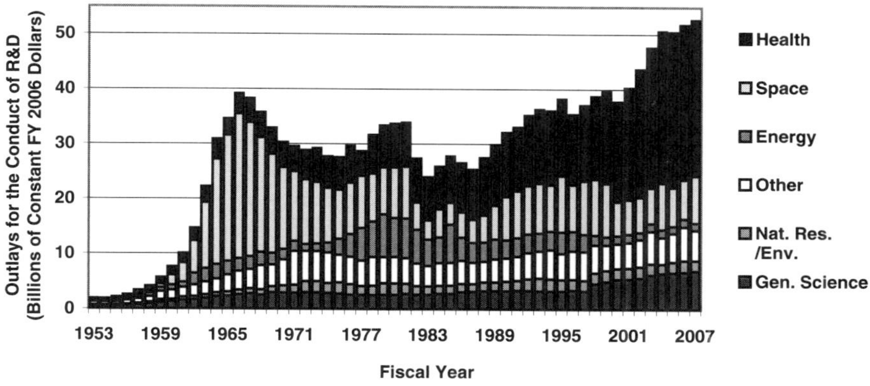

FIGURE 3 Trends in non-defense R&D by function, FY1953-2007.
NOTE: Some Energy programs shifted to General Science in FY1998.
SOURCE: American Association for the Advancement of Science, based on OMB historical tables in Budget of the United States Government FY2007. Constant dollar conversions based on GDP deflators. FY2007 is the President's request.

going to the National Institutes of Health, but has declined slightly since 2000. (See Figure 3.)

Unfavorable Trends in Spending on Science

A substantial amount of non-defense spending goes to space research, a "tiny" amount to energy, and a small amount to natural resources and environment. The general science category, aside from health spending, received little fiscal attention over time. "In support for the kind of work likely to push innovation," she said, "we're losing ground." R&D as a percentage of GDP has been declining since 1976.

Trends in R&D spending by business are characterized by a focus on development, rather than more risky basic research. In general, she said, "Research is driven by business needs, reliance on marketing insights, and a strong applied research orientation. Management makes a huge effort to maximize results from R&D." Overall business R&D funding was flat in 2003 and 2004, but rebounded in 2005.[7] This funding was found primarily in manufacturing, IT, and pharmaceuticals, and was dominated by a few large firms: Microsoft, Pfizer, Ford, General Motors, IBM, and Johnson & Johnson.[8]

---

[7]Data from the Industrial Research Institute.

[8]The Booz Allen Hamilton Global Innovation 1000: "Money Isn't Everything," *Strategy + Business*, Issue 41, Winter 2005.

Of the firms most admired for innovation—Apple Computer, Google, and United Health Group[9]—none was in the top 10 in research spending. Also not found in the top ten in research spending were the firms most admired for managing talent—General Electric, Proctor & Gamble, and Google.[10] This suggested, she said, that the most innovative firms were using the research of others rather than investing for the future.

One feature of the U.S. R&D landscape, she said, was the trend of state and local governments to recruit R&D organizations in the hope of increasing their economic growth. The State of Florida, for example, had recruited the Torrey Pines Institute for Molecular Studies, the Scripps Research Institute, and the Burnham Institute, promising them a total of about $1 billion in money, land, and other incentives. Other localities providing incentives to boost innovation in their own regions include:

- In Ohio, the Columbus 315 Research and Technology Corridor is a 10,000-acre development to be patterned after Research Triangle Park in North Carolina.
- In Iowa, $20 million has been allocated to the University of Iowa, Iowa State University, and the University of Northern Iowa—not for students but for economic development.
- In Michigan, $100 million has been granted to 61 companies to diversify the Michigan economy.

Other important players in the U.S. innovation system are found in the realm of the private foundations and non-profit organizations, which provide a significant amount of research support. Private foundations include the Howard Hughes Medical Institute and the American Chemical Society's Petroleum Research Fund, while non-profit R&D organizations include Battelle (whose motto is "The Business of Innovation") and many other significant entities.

The Scarcity of Seed Funding

Finally, she addressed the issue of early-stage funding for small R&D firms, which is a major emphasis in Flanders. In the United States, she said, capital is not always available when firms need it most.

She showed a chart indicating that VC funding for first-stage firms of relatively large size is still available. (See Figure 4.) Seed funding, however, for the smallest firms "is fast disappearing, and that's got to change." Startup funding, which once received almost as much VC funding as the first stage, had also

[9]*Fortune*, "America's Most Admired Companies 2006," 2006.
[10]Ibid.

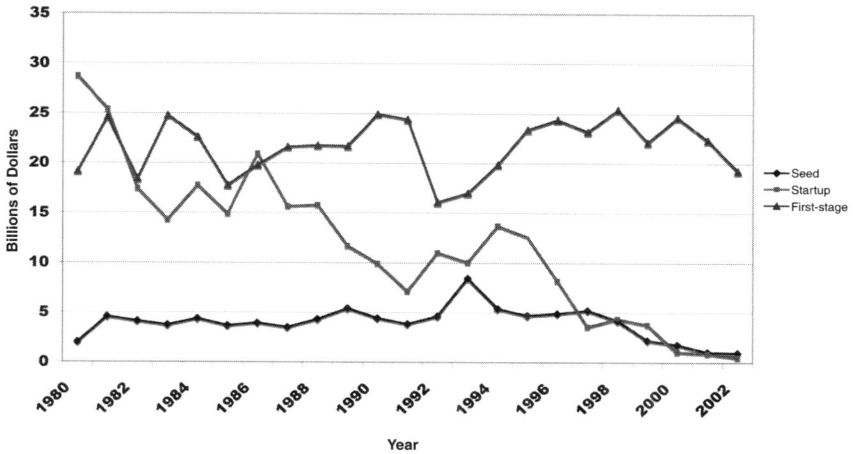

FIGURE 4 The collapse of U.S. seed and first-stage venture capital funding: dwindling high-risk investments.
SOURCE: National Science Board, *Science and Engineering Indicators 2004*, Arlington, VA: National Science Foundation, 2004.

declined,[11] as VC firms sought companies at more advanced stages that were likely to be less risky than startups. Without much federal support, she said, seed funding would have to come primarily from angels and state funds.

She summarized the challenges faced by modern nations, and the United States in particular, by dividing the issues of innovation into a series of three overlapping questions: (1) How do you get talent that does what you need it to do? (2) How do you raise sufficient support to give that talent opportunities? (3) How do you create an infrastructure capable of creating new and exciting things?

In response, she recommended actions in the same three categories with which she began her talk.

Talent

The United States needs strong emphasis on K-12 education at the national, state, and local levels. The universities must be recognized not only as providers of training and education of innovators, but also as engines of economic growth—without diluting the primary mission of education.

[11]National Science Board, *Science and Engineering Indicators 2004,* Arlington, VA: National Science Foundation.

Investment

The nation needs broad initiatives to provide new investments in research, models to stimulate private-sector innovation, and R&D tax credits. There must be more early-stage funding models for small firms, including local, state, and angel funding. Business must renew its investment in R&D, with revised management structures, to maximize the total innovation chain.

Infrastructure

We must move from the discussion stage to the action stage to focus on metrics that measure innovation strategies. We need new organizational models to accommodate interdisciplinary R&D and external partnerships, as well as support for and integration of the manufacturing and service sectors.

Discussion

A questioner asked Dr. Good's opinion of the competition between states for new industrial plants. She said that competition for R&D facilities probably did no harm, if the facility was producing something new, but she deplored the enormous expense of tax money spent by some states to gain a straight manufacturing plant that might or might not repay the investment.

GLOBAL COMPETITION, CORPORATE POLICY, AND NATIONAL INTEREST

Mark B. Myers
Xerox Corporation (retired)

Dr. Myers began by emphasizing the point that global innovation occurs within a vast but interactive system, so that no single element is sufficient to dominate it. In some ways, that system had been U.S.-centered for many years, although the number of competing participants and dispersal of resources were growing rapidly.

The United States has traditionally deployed its own innovation resources very effectively, he said, supporting a broad portfolio of R&D in basic science and technology. Its funding pattern was part of the general national strategy of investing broadly—of supporting a diverse portfolio of pre-competitive technologies. It maintained an open R&D system in which results are published and freely available and depended on spillovers, mainly from the large proportion of basic science performed by the Department of Defense, to energize the private sector. The private sector created and supported technical innovations through a combination of venture capital, large corporate research laboratories, and the activities of startup firms. Schools of engineering and medicine provided sources of spin-outs and applied generally balanced policies of IP protection.

Major Changes in the 1990s

Through the 1990s, the United States performed well, raising its share of the high-tech global market, increasing its private funding of R&D. Some of the features of U.S. performance then began to change, including more privatization of information, an increase in patents filed, and reduced numbers of U.S. scientific pubs. As economy shifted away from manufacturing and toward service activities, the large corporate research labs began to downsize and even disappear. The importance of the research universities rose as they took over the role of the corporate labs, and the "perimeter" of the university, which began to include considerable industrial activity, started to become industrial R&D centers. Monopoly powers disappeared from industry in the 1980s and 1990s, causing great shifts, including the replacement of science-driven R&D by market-driven R&D. The activity of venture capital firms rose rapidly in the 1990s and then fell just as quickly. The globalization of R&D, which no one had foreseen, gave the global innovation system a newly dispersed structure.

The 1990s also saw large technical transformations. The first was "Moore's Law," which described the regular and rapid increase in microprocessor capacity and the parallel revolution in the speed of product development. This new speed meant that the technology underlying most business models was now constantly under attack and that vertically structured organizations in the PC industry had to become horizontal quickly. IBM barely survived this revolution, while many others—including DEC, Sperry-Univac, and Wang—did not. The shape of enterprises today is harder to define, with multiple centers and virtual connections.

The Transformation of Corporations

Other key transformations were brought about by wave division multiplexing, optical networks, and the Internet. Businesses became networked enterprises: flat, virtual, dependent on outward engagement, with competency centers arranged globally. The very definition of companies became blurred. Is Dell a computer company? he asked. Dell spends less than 1 percent of revenues on R&D, while the computer industry as a whole spends 12-15 percent. Is Dell a technology company at all? The answer, he said, is that Dell is a technology company in the same way Wal-Mart is a technology company. Both have focused their innovation activities on the supply chain, where the R&D available to them is very sophisticated. The technological revolution, he said, had enabled a different kind of innovation in the way firms are designed at the global level. Traditional kinds of competition, once defined in terms of a global place, can become irrelevant because of the nature of the technology. He cited the example of photography, a field in which Kodak and Fuji Photo had long competed: Neither firm today makes a profit in photography, which was based on a silver halide technology. The firms making a profit in photography today are Canon, Epson, and H-P, who sell digital cameras and whose strength was in information technology rather than traditional photography. A combination of geographic dislocation and technological dislocation now defines the competitive space, he said, in which a new entity—the global company—is dominant.

As a result, large firms increasingly have to play globally or they will not survive. The global corporation today:

- Has a worldwide customer base.
- Does R&D that is market driven: the firm needs science, but does not invest in it.
- Has a new balance of global and national perspectives.
- Seeks the best talent wherever it is available.
- Finds the technology it needs through investment, partnerships, and acquisitions.
- Forms dynamic partnerships, and makes acquisitions aggressively.
- Makes use of networked and open innovation.
- Depends on international standards.
- Maintains a relentless drive for productivity.
- Emphasizes risk management: Can we afford to spend $1 billion on a new drug and not get one?
- Needs partners. If you depend on a worldwide supply chain, you need help if that chain is disrupted.

Thorny Issues for Innovation

Going forward, Dr. Myers saw a series of thorny issues for the global innovation system. First, each nation must have policies that address the globalization and dynamic linkages of modern firms. As universities develop their own "innovation perimeters," where entrepreneurship is the focus, they must grapple with effects on the primary missions of education and research. Governments, industries, universities, and others must agree on how to fund the "knowledge commons" on which innovation depends. Nations must better deal with workforce capabilities and location as migrations increase across the world. Industries must learn to deal with the different interests of small and large firms, as more growth occurs by acquisition. Few small firms, if attractive, will grow to be large, creating a particular problem for small countries that have difficulty growing large firms. Governments must harmonize their IP policies to sustain the exchange of knowledge.

Given such a list of complex issues, Dr. Myers confessed that he was somewhat pessimistic about the United States' ability to respond quickly with innovation policies appropriate to this global age. He concluded that unless national leaders can make a persuasive case for policies that are necessary to ensure long-term global competitiveness, they may be forgotten amid more obvious but short-term priorities.

Discussion

Dependence on Old Research

Dr. Spencer noted that many innovations driving the economy in recent years, such as the transistor, depended on basic science done 50 years ago. In the face of declining public and industry support for basic science, he asked, where would the inventions for continued innovation come from? Dr. Myers agreed with this characterization, and confessed that he had no answer to the dilemma. He found "disturbing" the long-term decline of federal investments in engineering sciences at universities which, combined with the lack of attractiveness of engineering for the American native population and the restrictions of immigration policy, "may cause severe problems. We may need an 'innovation shock'," he said, "as we had from Sputnik."

The Importance of Small Advances

Professor Soete commented that he could imagine a future of continuous technological expansion based on old technologies, as in the field of medical diagnostics, where the research is "independent, individualized. You're opening up dramatic new areas of discovery, medical areas which are small but are being perceived as extremely useful for the social welfare." He said that a lot of research at "the bottom of the pyramid" was fascinating because it "challenged the innovation trajectory as we know it, adding features to products cheaply that really help." He cited the example of wood stoves that are 100 times as efficient as older models, but very cheap.

Professor Good agreed with Dr. Spencer's comment that innovations today are "built on a pool of science 50 years old. If we don't replenish the pool, there will be no fish." Innovations in medical imaging, for example, such as the MRI and PET scanners, are based on fundamental but old physics. She said that the United States has probably lost its lead in the high-energy physics that led to such instruments, and U.S. high-energy physicists now go to CERN in Europe. High-energy physics had also led to modern cryogenic engineering, which has also declined in the United States. "Our federal R&D is not keeping up," she said. "And the private sector is not going to do it." She added that strong basic research is particularly important for the United States, which decided after World War II to link university research to its training of graduate students in technical fields. Declines in research automatically weaken the training component of that effort.

The European Focus on Jobs

A questioner asked about the EU strategy set up in Lisbon in 2000 with the goals of a knowledge-based economy and stable jobs for people. He said that jobs should be the main goal, but that the new strategies said nothing about this—only about how to spend 3 percent of GDP on R&D by 2010. He questioned the value of scientific papers and invited speakers, and asked what is being done to provide the many new jobs that had been promised for Belgium. Professor Soete acknowledged that at Lisbon the ministers of employment were talking about such issues as employment targets and the participation of women, while the ministers of S&T were talking about knowledge, but the two elements were never linked. He said that his response as an economist was that the ultimate aim of a knowledge base is increased welfare—a concept broader than GDP.

Dr. Myers added that the time lag between the discovery and application of knowledge compounded the problem for policymakers. At Xerox, he said, the lag between the investment in a research project and the point of peak revenue was 8 years. In pharmaceuticals, the lag is about 13 years. "Most systems are not set up to measure that," he said, "especially when it needs to satisfy political needs."

Dr. Good noted that good retrospective studies had been done on the value of scientific knowledge, and that economists agreed that more than 50 percent of GDP in the United States since 1950 had been generated by technological inventions. She said that evaluations of investment in R&D must be done on that basis to be meaningful.

Dr. Wessner noted "an important political point," saying that the outcomes of research are not linear, take time to appear, and are often unexpected. He said that the considerable value of IMEC was measurable in many ways, including not only the scientific output but also the employment generated and funds spent by visitors. He also returned to the 3 percent issue, noting that this goal had been discussed in Europe for 6 years and yet countries were still not making serious efforts to reach it. In addition, it seemed unlikely that institutions and regions had the R&D capacity to absorb such a large increase in funding as rapidly and productively as hoped. Finally, relieving unemployment was a complex and general problem that would require many kinds of structural changes, including opening markets.

The Complex Route to More Jobs

Professor Soete agreed that Europe urgently needed structural economic changes before it could expect better employment and living conditions. He granted that more technical training would play a small role in this. The demand for workers in the health care sector, for example, is anticipated to exceed the total output of most countries' educational systems. And the number of technology-related employees at the universities in Flanders had increased 70 percent over the past 15 years, so the region's high-tech policies had already generated additional employment. But this was only a small fraction of what could be achieved by much-needed fiscal measures, such as the reduction of social security payments.

These necessary macro-economic changes, he said, had nothing to do with the high-skill jobs being discussed at the conference. He attributed employment imbalances to general mismanagement of labor markets, the failure to open up more markets, and slow progress in using new technical knowledge to generate employment. Europe did not need a policy that puts "everybody into the labor market, no matter what they do," but "a much more strategic policy of increasing the knowledge intensity of economic activity."

Dr. Myers closed the discussion by mentioning the "Solow paradox," the discovery that when U.S. firms first invested in information technology, they saw no increase in productivity. The problem was that new technology was being applied to existing work processes. His company found that productivity increased only when work procedures, including the production floor, were totally redesigned to fit the new IT. Any discussion of jobs, he said, needed to include a discussion of productivity, both of which are important for secure economic performance.

Keynote Address

Fientje Moerman
Vice Minister-President of the Flemish Government
Minister for Economy, Enterprise, Science, Innovation, and Foreign Trade

Mrs. Moerman said she would give an overview of what Flanders had been doing and planned to do with its research and innovation policy in Flanders. She noted that the current symposium grew out of a visit to Washington a year earlier, when she visited the U.S. National Academy of Sciences. She thanked Dr. Wessner and his team for their enthusiastic support and commitment to the event.

Flanders' Investment in Innovation Policy

Investing in knowledge and innovation is crucial to sustainable growth, she began. At the Lisbon European Council of 2000, the EU heads of state expressed their desire for the EU "to become the most competitive and dynamic knowledge-based economy in the world, capable of sustainable economic growth with more and better jobs and greater social cohesion." The Barcelona European Council set a target of 3 percent of GDP to be spent on research and development by 2010, with two-thirds coming from industry and one-third from government.

Before that, she said, the government of Flanders was critically aware of the importance of R&D to its economy and welfare. The transformation of Belgium into a federal state and the devolution of nearly all powers with regard to education and R&D to regional public bodies marked the start of Flanders' strategies. At that moment Flanders' public expenditures on R&D were far below the European average. Since 1995, however, successive governments had more than doubled public funding for research and technological innovation to a level well above the EU average.

Key Mechanisms of the Innovation Strategy

She noted that larger budgets are necessary but insufficient for successful science and innovation policy; spending the resources in the right way is equally important. Taking into account best practices of the international environment, particularly from the European context, the Flemish government had developed an innovation strategy creating appropriate funding mechanisms and instruments to monitor and evaluate its policy on a regular basis. She summarized the main characteristics as follows:

1. Maintain a double budgetary balance—part for academic basic research and technology innovation, and part for higher education institutes and industry.
2. Adopt a bottom-up approach. Apart from the strategic research centers, the government has set few thematic priorities, funding instead projects proposed by the researchers themselves.
3. Give universities and interuniversity research institutes, such as IMEC, a large degree of autonomy. The Flemish government sets out annual block grants, long-term performance targets, and long-term management agreements. Performance-based funding is the key.

Addressing the "Innovation Paradox"

In 2003, the Flemish government concluded an Innovation Pact with key players from academia and industry to reach the 3 percent Barcelona target. The Flemish Science Policy Council (VRWB) was designated to monitor the execution of the pact, using 11 key indicators. The first findings, published in 2005, were that Flanders is characterized by an average innovation profile and was insufficiently able to transfer excellent (academic) research findings into innovative products or added societal value—the "innovation paradox" that afflicts most European and other countries. Just a few, mostly international, companies accounted for all industrial research in Flanders, leaving the economy vulnerable to external events and corporate decisions.

The challenge was to reduce the innovation paradox, which meant reducing a traditional culture gap between industry and academia. Academic researchers had long felt that working in industry corrupts the academic career, diminishes publication output, and restrains academic freedom. At the same time, industry described a structural mismatch between the research agenda of academia and the research needs of industry.

A recent study by the Catholic University of Leuven (K.U.Leuven) sent more positive signals about the interaction between industry and academia. It suggests that the gap between industry and academia is shrinking. First, the study estimated that in 2005 about 10 percent of all R&D in Flanders was performed by academic-industry partnerships. This was in line with statistics in the Third S&T Indicators Report by the European Commission (2003), which found that the relative share of industry in R&D expenditure in higher education in Belgium was 10.9 percent, far above EU (6.9 percent) and U.S. (6.3 percent) levels. Second, from 1991 to 2004, universities and public research institutes created 101 spin-off companies, 54 of them in the past 5 years. Third, research teams that work closely with industry also perform well in basic research.

Reducing the Gap Between Academia and Industry

She then turned to measures the government is taking to help reduce the gap between academia and industry. First, in 2004, the government established the Industrial Research Fund (IOF) at the universities, with an annual budget of about €11 million. This is distributed to universities on the basis of performance-based parameters, such as number of spin-offs created, number of patent applications, volume of industrial contract research, and budgetary share in the European Framework Programme. The IOF allows for hiring postdoctoral staff, who concentrate on research findings that show great near-term potential for market applications. This group of researchers is evaluated on the basis of their applications-oriented performance. In the near future, the IOF will also allow universities to fund strategic basic research projects. It also allows every university and associated college of higher education ("hogescholen") to pursue its own policy of strategic applications-oriented research. The aim is to stimulate industry-oriented research and support excellent research groups in industry-relevant areas by giving them longer-term funding.

Second, the government has set up interface units, or technology transfer offices, at universities. The budgets for these TTOs will double over 2 years through 2007. The goals is to help offices toward further professionalization of staff and services and help them include university-associated colleges of higher education.

In addition to the IOF and the TTOs, the government is exploring additional initiatives, including intersector mobility of researchers and the use of predictive methods to assess the potential economic impact of technologies. Intersector mobility between academia and industry is paramount for the exchange of knowledge and methodology, to refine the research agenda, and familiarize young researchers with the industrial working environment, where more and more will find employment. The main existing fellowship scheme for PhD students is managed by the IWT, the Flemish Institute for the Promotion of Innovation by Science and Technology. Fellows submit an applied research proposal, typically for 4 years, allowing them time to obtain their PhD. The IWT also runs a limited postdoctoral program that funds researchers planning to set up their own spin-off company. These programs focus on applied research, but have not yet reached their goal of creating intersector mobility in the sense of moving people between companies and university labs in both directions.

Risks and Benefits of Foresight Exercises

With regard to foresight, the major exercise being undertaken by the Flemish Policy Research Council (VRWB) is trying to identify major S&T areas of the future, taking into account their current economic capability, research potential, links with current international trends, and potential for future growth. The VRWB came up with six clusters: (1) transport; (2) ICT and health care services; (3) health care and treatment; (4) new materials, nanotech, and the processing industry; (5) ICT for socio-economic innovation; and (6) energy and environment for the service sector and processing industry.

This foresight exercise contrasted with the use of thematic priorities to determine funding channels, which she criticized as being "top-down" and not leaving enough "breathing space" for bottom-up initiatives and smaller research players. She preferred an open, no-strings-attached strategy, which invites research proposals defined by the industrial and academic communities themselves, using peer review to the extent possible.

The results of the VRWB foresight exercise, she said, might be useful for choosing among new large-scale projects of research consortia. She mentioned the "clusters of competence," bottom-up initiatives by industry to create a critical knowledge platform in their sector. Open innovation is the underlying principle, with knowledge made available to all participants. Research is done in close collaboration with multiple industrial partners so that costs and risks are shared. About ten areas of competence are now funded, in areas such as logistics, food, mechatronics, geographical information systems, product development, and industrial design.

Non-technological Aspects of Innovation

A second major policy challenge, she said, was to broaden the concept of technological innovation to include its non-technology dimensions. Until two and a half years ago, Flemish innovation policy had only targeted the technological dimension of innovation. There was a growing awareness, however, that innovation also touches on management, public and private governance, labor market organization, public procurement, and design. The challenge is to develop policy elements that cover these elements.

One policy priority is to "mainstream" innovation so that it becomes a horizontal dimension in all policy fields. The Flemish Innovation Plan, approved in 2005, specified nine main lines of action cutting through all sectors. Each year a *status quaestionis* of achievements will be produced. Among those designed to date:

- With the Minister of Environment, she established the Environmental Innovation Platform (MIP), which brings together all relevant stakeholders and acts as a catalyst for innovation in a domain in which Flanders has much "unclustered" knowledge and expertise.

- For FY2007, the government set aside a budget for "media innovation," another cross-cutting area with huge economic potential.
- A roundtable was organized with specific industrial sectors, such as the life sciences and chemical industries. The group has listed innovation deficiencies and obstacles to economic growth and drawn up action plans which are being executed and closely monitored. The roundtable, she said, had allowed for the frank exchange of views between the government, labor unions, companies, and research institutes, and offered practical solutions.

Finally, she noted the need for strategic, international intelligence. Two factors make it imperative that governments join forces across borders: (1) the growing rate of globalization, and (2) complexities presented by an open innovation system in which governments no longer have sufficient instruments to create an adequate policy mix. Flanders needs to enhance the mutual understanding of its science and innovation systems, she said, both nationally and internationally.

Importance of Networking for a Small Region

In Flanders, science and innovation policy preparation is the main task of the recently created Department of Economy, Science and Innovation. However, high-quality, evidence-based policy can be prepared only by bringing together people who know both theory and practice on daily basis, such as the universities, junior colleges, and companies. Desk study and field work, she said, have to be combined. One challenge is to bolster the pool of S&I management knowledge in Flanders and to network the agencies and organizations that carry out science and innovation analyses, often on an ad hoc basis. The Flemish research landscape is so small, she said, and its capacity so limited, that only a networked approach can yield efficient results. It makes no sense for small and often isolated study units at various organizations to be unaware of each other's activities. A networked approach is one of her policy priorities for the coming months and years. Another is increasing first-hand field knowledge of those responsible for policy preparation. She plans a mobility program to allow for the temporary exchange of staff members between administrations, funding agencies, universities, public research institutes, schools of higher education, and companies. Such a program, she said, will make participants aware of the peculiarities of other, often unknown, environments. It should also reduce the number of superfluous rules when designing new programs.

Innovation is a global challenge, she concluded. To innovate is the key to survival, economic growth, and social welfare. On both sides of the Atlantic are the assets of excellent basic science, internationally minded young scientists, and state-of-the-art research equipment. The growing complexities of open innovation and the increasing challenge of eliminating the innovation paradox are strong drivers for great mutual understanding and exchange of views. She proclaimed this Flanders-U.S. "innovation dialogue" a good start and thanked the organizers and host.

Discussion

Dr. Spencer said that the amount of foreign direct investment in the Flanders area was impressive, and asked whether it could be enhanced by having faculty from American and European universities spend time in Flanders. Mrs. Moerman said Flanders had developed a program called Odyssey to re-attract Flemish scientists who have emigrated abroad to do their research, as well as some researchers from across the world. The universities have a large degree of autonomy, she said, so they were free to attract academic researchers from various countries.

A questioner asked how IP regulations affected innovation in Flanders. Mrs. Moerman said that there was indeed a culture problem. The universities traditionally receive an amount for seed funds, and the Flemish government had doubled that amount and set up a general scheme for dividing profits earned from IP between the university and the inventor, along with a fair tax measure. She acknowledged that IPR questions continued to present complications, especially where different patenting systems were involved.

Cooperative Research and Global Competition in Semiconductors

Moderator:
Peter Spyns
Department of Economy, Science, and Innovation
The Flemish Government

CURRENT TRENDS: A U.S. INDUSTRY PERSPECTIVE

George Scalise
Semiconductor Industry Association

Mr. Scalise gave an upbeat assessment of semiconductor market trends, calling it "a great market today," after 3 solid years of growth. For the current year, he said, the market was forecast to grow by nearly 10 percent, having already grown 8 percent in 7 months. The market was being driven most powerfully by consumer demand for products such as cell phones, digital cameras, digital TV, personal computers, and MP3 players—which accounted for more than 50 percent of demand worldwide. The industrial sector share had dropped slightly below 50 percent. The data going forward, he said, suggested compounded IT revenue growth of about 10 percent "as long as the world economy continues to do well," especially India and China.

Of product areas, MOS logic was by far the most important. Flash memory was replacing rotating memory, he said, a trend that would accelerate, and analog devices were being "pulled along." Optoelectronics, with growing use in sensors, was becoming a major contributor, while the fastest growing segment was digital signal processing.

New Product Trends

New products would continue to be cheaper and more powerful, continuing a long trend. Comparing the personal computer of last year with a PC in 1995, he cited a storage capacity 100 times higher and an overall price decline of 98 percent. "The overall functionality/cost equation makes it incredibly cheap to buy a PC now," he said, "and I don't see any reason why that won't continue."

Semiconductor technology had entered the nanometer range already, bringing a "whole host of challenges." In about a decade, he said, the continual shrinking of semiconductors would bring the industry up against physical barriers—power dissipation limits, technological limits, and economic limits—that "may slow us down a little bit." He cited heat dissipation as a particular problem. But he predicted that the industry was on the right track with both design solutions and process technologies to continues its progress.

What Lies Beyond CMOS?

Another particular challenge is to find the next generation switch beyond CMOS,[12] which he said would probably be required in 10 or 15 years. None of a half-dozen current alternatives to the CMOS logic switch are close to being useful alternatives.

With regard to end uses for semiconductors, said Mr. Scalise, product rotations were being driven by the consumer now, and product cycles were quickening in response to consumer demand. Cell phone cycles, for example, had dropped in the last few years from 28 to 16 months. Prices had come down, functions had risen, and that trend would continue. To stay in business, companies had to be closely tuned in to what consumers want, and to be in the best position to meet that demand at the right time.

As the size of transistors continued to shrink, he said, the industry will have "multi-dimensional innovation requirements." Today, they are still using equivalent scaling to follow the pace of Moore's Law. That is, there continue to be new materials and device structures, but still within the existing CMOS scaling environment. Something fundamentally new will be required as the transistor passes below 32nm[13] and power dissipation issues become acute. Many experts, he said, think that at least part of the problem can be overcome with atomic layer deposition techniques.

---

[12]CMOS, or complementary metal-oxide-semiconductor, is the dominant technology mode for digital integrated circuits. The CMOS transistor was invented in 1963.

[13]The nanometer term describes the size of the smallest feature that can be manufactured on a single chip. There are about three to six atoms in a nanometer, depending on the type of atom. Reducing the size of the features enables smaller, more energy efficient and powerful chips.

A "Big Picture" of Semiconductor Research

He offered a "big picture" of what the industry is doing to address such challenges. SIA divides its initiatives into competitive (1-3 years), pre-competitive (3-8 years), long-term (8-14 years), and exploratory (15+ years) R&D programs. SEMATECH plays a central role in developing tools and infrastructure, primarily in the pre-competitive stage, and the Advanced Transistor Development Facility (ATDF)[14] makes its fabrication capabilities available to SEMATECH and others.

In addition to SEMATECH, the industry benefits from the Semiconductor Research Corporation (SRC), a group of chip makers and about 100 universities. The SRC's Focus Center Research Program (FCRP) addresses the industry's "most intractable problems," such as the physical limits of silicon, increasing product complexity, shrinking design cycles, reduced long-range research budgets, and the dwindling supply of qualified engineers. Its research program involves 5 centers, about 35 universities, 200 faculty, and 400 graduate students to "drive the technology forward and bring out new young talent."

Another SRC program is the Nanoelectronics Research Initiative. Its research objective is to explore ideas and demonstrate proof of concept for a new logic device by 2020. Industry participants include AMD, Freescale, IBM, Intel, Micron, and Texas Instruments. "Structurally," he said, "we have all the components covered."

Mr. Scalise concluded by emphasizing that nanotechnology innovation requires the partnership of government (deep expertise in fundamental research), industry (knowledge of technology transfer, road mapping, and the path to commercialization), and academia ("out-of-the-box" thinking and new ideas). The NRI now has only about $8 million in annual funding, but during the next 3 to 5 years will scale up to $200 million or so. "We'll need that to meet the challenges at the nano level."[15]

Discussion

Dr. Wessner asked whether the federal government was prepared to make the large expected investment in nanotechnology, and whether the effort would be national or international. Mr. Scalise said that the NSF understands the need, but that the next stage, which would require passage of the kind of legislation

---

[14]ATDF is an independent subsidiary of SEMATECH's R&D wafer fab and associated analytical laboratories. According to SEMATECH President and CEO Mike Polcari, "While the SEMATECH consortium continues to focus on our core business of building industry infrastructure in lithography, materials, and manufacturing, the new company represents a complementary effort to meet the more targeted R&D needs of individual companies and universities."

[15]A large coalition is also needed to pay for semiconductor R&D costs, which are increasing almost twice as fast as revenues, according to ATDF.

that launched SEMATECH, had not been worked out. He referred to the three pillars of the President's American Competitiveness Initiative (ACI), and said that the basic research would be ready "when the time comes." The other two challenges of the ACI are larger. One is to ensure a skilled workforce by that talented students continue to come to the United States and stay here to work along with improving K-12 math and science education. The second is to choose to compete for investment in design and manufacturing projects—the focus of competitors around the world who use incentives and changes in tax policy.

CHINA'S INNOVATION POLICIES

Alan Wm. Wolff
Dewey Ballantine LLP

Mr. Wolff, who said his work on behalf of the semiconductor industry had taken him to China for the past 10 years, opened with a picture of a billboard near the entrance to the city of Suzhou. On the billboard was written: "Development is an immutable truth." That priority has been fulfilled, and has involved very heavy technology-based development. That commitment was described by Jiang Zemin, then General Secretary of the Communist Party of the China Central Committee, who said in 1999: "In today's world, the core of each country's competitive strength is intellectual innovation, technological innovation, and high-tech industrialization."

This philosophy is pervasive. In contrast with Western leaders' brief comments about innovation in statements of their priorities, said Mr. Wolff, it is a theme in every Chinese leader's talks. The objective of the strategy is to progress from imitation to production to creating indigenous technology products: ". . . to move from 'Made in China' to 'Made by China.' It is an objective second to none."

A National Policy of Investment in Technology

Among the most significant basic documents describing Chinese thinking is the 15-year "Medium and Long-Term Program on Science and Technology Development (2006-2020)." This program specifies intensive investments in crucial high-technology products, using policy tools to reward technologies made at home. China plans to increase R&D spending to 2.5 percent of GDP by 2010, which will about equal that of the United States. The State Council of the PRC has issued long lists of technology-based objectives, from core electronic components and new drugs to manned space flight and lunar exploration. The country has published "guiding opinions" (99 altogether) to move China in a technological direction, advising on such topics as corporate bonds, startup investment funds, debt financing, development zones, and venture capital. According to the 11th Five Year

Plan, pieces of which were just emerging, ". . . [China] will promote development by relying on enhancing independent innovation capability, as a national strategy shift in economic growth from relying on the input of capital materials to relying on scientific and technological advancement and human resources."[16]

The national IPR strategy, said Mr. Wolff, was to use measures to improve national competitiveness—even if they push against global standards. In the words of one official: ". . . [we shall] abide by international principles and meet the lowest standards of the WTO. . . ." and "[we shall] not only encourage self-innovation, but also encourage absorption, consumption, and innovation of introduced technologies."[17]

Measures to Encourage Technology Transfer

China has also adopted powerful tools to encourage technology transfer and encourage foreign investment in R&D. One is to exempt from sales tax income earned from the transfer of technology developed exclusively through foreign direct investment in R&D. Another is to give foreign R&D investors with rising development expenses a 50 percent discount in corporate income tax.[18] A third is to design procurement regulations that favor domestic products.[19] China's import policy is similarly designed to help China by "watching what comes in and absorbing it." One ministry recommends increasing "the investment in assimilation and absorption" of imported technologies to "gradually establish a market-oriented system of" technology imports and innovation.[20]

Other important policies, he said, attempt to guide development. A key one is an antimonopoly policy that aims to "prevent vicious competition in the industries, which if used in a discriminating fashion, could impair foreign investment which has been central to China's drive to innovate." In addition, to promote investment, the government provides relief from "social responsibilities." Local government authorities have set aside billions of dollars to build semiconductor fabs for Chinese companies.

Incubation parks are important to the national innovation strategy, and they had, as of 2005, according to Chinese government sources, attracted some thousands of companies. The Tianjin Binhai New Area for biotechnology is twice

[16]Ma Kai, Minister, National Development and Reform Commission, 2006.

[17]Lu Wei, Deputy Director General, Technical Economic Department, Development Research Center of the State Council, 2005. Foreign R&D investors with development expenses at least 10 percent greater than previous year expenses are entitled to a 50 percent discount in total technological development expenses in the current year corporate income tax.

[18]Guogong Long, "China's Policies on FDI: Does Foreign Direct Investment Promote Development?" May 2005.

[19]*Outline of the National Medium- and Long-term Program on Scientific and Technological Development, 2006-2020,* State Council of the Peoples' Republic of China, 2006.

[20]Ministry of Commerce, 2006.

the size of the Flemish province of Brabant, said Mr. Wolff, with investments by 69 companies of the Global 500. Shanghai Zhangjiang, a 16-square-kilometer high-tech park, is viewed as China's "Silicon Valley"—and "Pharmaceutical Valley" as well. Its stated goal is "to form a perfect high-tech innovation chain." It now hosts 42 foreign companies, including Roche, GlaxoSmithKline, and Medtronic, as well as 70 fabless companies and three semiconductor foundries.

China, unlike Japan, has encouraged foreign direct investment as a key component of its innovation policy. It has also begun to improve IP protection, and has created incentives for indigenous patenting. The number of patents granted had risen 88 percent from 2001 to 2005, and "high-tech" exports have grown rapidly.

Conditions that May Hold Innovation Back

At the same time, some conditions hold innovation back. Among them, he said, are state planning, the active participation of the Communist Party, and a considerable level of corruption. In some universities, the engineering curriculum requires several hours a week of Marxist philosophy, a distraction from China's economic goals. Other drawbacks of China's top-down system are instances of "techno-nationalism," he said, including attempts at forced technology transfer, favoring domestic companies through national standards requirements, managed trade, misguided industrial policy, and misallocation of capital.

While the output of the educational system is massive, some observers have also expressed doubts about the quality of the S&T workforce. According to studies by Duke University, McKinsey, and Cao and Simon, China's educational system is outdated in emphasizing depth over breadth, a quantitative over a qualitative focus, and neglecting to nurture creativity. Mr. Wolff said that these features tend to produce graduates who do not meet the hiring needs of major western companies.

To the country's credit, however, its leaders are well aware of some of the defects of their system, said Mr. Wolff. For example, the Ministry of Science and Technology lists conditions that hold back the development of sound IPR protection:

- No recent history of private property
- No history of a culture of IPR
- A share of world patents that is still very low
- Patent quality that is low
- Almost complete lack of patent ownership by Chinese firms

IPR Abuses

Mr. Wolff mentioned the issue of IPR abuses. According to the State Intellectual Property Office, he said, the reason the state does not crack down on counterfeiting is that people would not be able to afford the resulting prices for products. Other deficiencies in IPR protection may be that certain features jeopardize the stated goal of foreign direct investment, including a tendency by Chinese partners to withhold core technologies, limit technology transfer to the routine, hold back key IPR components, and achieve only limited synergies with other Chinese companies.

In summary, Mr. Wolff said that China's innovation system is a work in progress that continues to depend on external input. It is probably held back by the dominant role of the state, he suggested, concluding with a question: Can a government intervene in the market as deeply as China's does and have a market economy that maximizes innovation?

Discussion

Dr. Spencer asked how China determines when an invention is Chinese, as opposed to European or American. Mr. Wolff said local innovation is still the exception, but the steady inflow of repatriated engineers from around the world is likely to raise the level of local innovation.

INTRODUCTION TO IMEC

Anton de Proft
IMEC

Dr. de Proft, chairman of IMEC, began by commenting about its success, which he attributed at least partly to a policy that is "kind of hands-on but from a distance." The government asks for a 5-year program plan, which is followed by in-depth evaluation and the adoption of performance indicators, and then another 5-year program. This, he said, is intended to avoid micro-management.

The Goal of Being a Worldwide Center

The goal of IMEC, he said, was to be a "worldwide center of excellence" that focuses on exploratory work with a significant impact on industry. Since its founding in 1984, its staff had grown from 70 to 1500, and at the time of the workshop it had about 500 corporate partners. It is subsidized by the Flanders government, contributing 17.8 percent of the budget in 2005.

About 22 percent of those working at IMEC are industrial residents who live in Leuven for a year or more, and 14 percent are from academia in Flanders. More than 50 countries are represented, including France, the Netherlands, Germany, Japan, Korea, China, the United States, and many others.

IMEC's basic technology platform, he said, is nanotechnology and its overall mission is "making things smaller, better performing and allowing to address

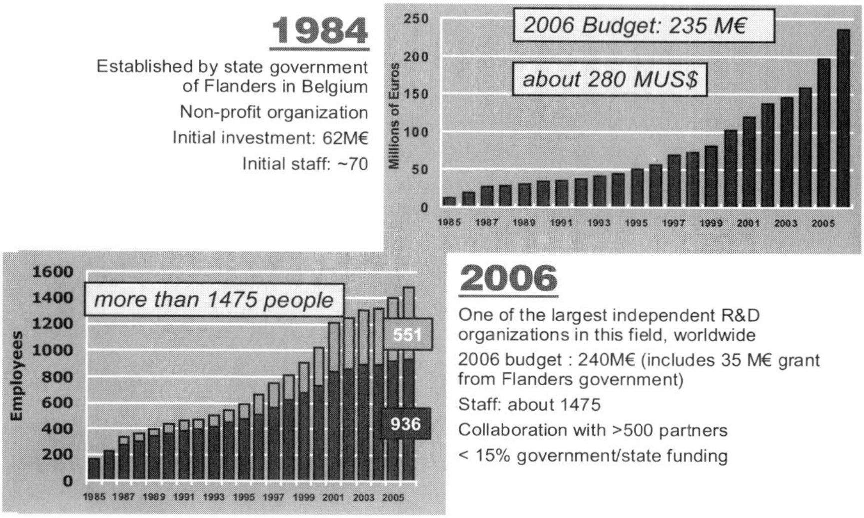

FIGURE 5 IMEC: More than two decades of open innovation.

a wider range of densely integrated functionalities. We explore how to move into a multidisciplinary world where we must do many things simultaneously, leveraging on our strongly deployed research infrastructure and wide range of competencies."

Challenges for the Semiconductor Field

He summarized some of the challenges for the semiconductor field in coming years. First was the cost of semiconductor R&D, which is increasing by about 12 percent annually. "This wasn't a big deal when revenues were going up faster," he said. "But 10 years ago revenue growth slowed, and the consensus now is for roughly 6.5 percent revenue growth. The only way to keep the R&D budget under control is by sharing costs and allowing access to external R&D."

Within the product life cycle, IMEC positions itself at the non-competitive stages, "right after university work," where joint research is appropriate and more and more a necessity (must-have technology platforms). It acts as a "transformer" between academia and industry, providing both greater focus for universities and basic insights and solutions for industrial partners.

The overall budget for IMEC was about €235 million for 2006. In 2005, the largest portion of revenues (49 percent) came from both core and non-core partners. Core partners included Intel, NXP, Texas Instruments, STMicroelectronics, Infineon, Micron Technologies, Samsung, Panasonic, and Taiwan Semiconductor.

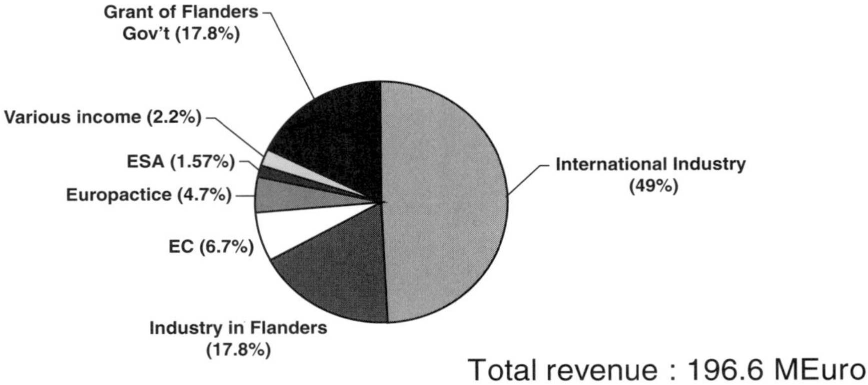

FIGURE 6 Sources of revenue in 2005.

IMEC also has strategic partnerships with about a dozen equipment suppliers and many "non-core" partners. (See Figure 6.) The Flemish government and industry in Flanders each provided about €35 million, with smaller amounts from the European Commission, ESA, and others. He added that while IMEC is careful about claiming to be the biggest anything, "I think it's fair to say that this is the world's largest industry commitment to semiconductor research in partnership."

He showed a drawing of the IMEC campus, which holds both its original 200mm pilot line in Clean Room 1 and the new 300mm pilot line in Clean Room 2, which have total clean room space of about 8,000 square meters. About 45 tools had been installed in the 300mm room; the equipment arrived in August 2006.

Toward an Interdisciplinary Future

In the future, he said, IMEC would continue bringing complementary and interdisciplinary expertise under the same roof geared towards an increased speed of innovation. One example is the Neuro-electronics Convergence Laboratory. The different expertise included institutes of medicine (Leuven Faculty of Medicine), biology (the VIB), nano/micro electronic (IMEC), and chemistry (IMEC). All of them share facilities, space and expertise in cross-disciplinary projects at the micro- and nanoscales. Part of their philosophy was to look at many different technology options and the many trade-off aspects, because no one could tell in advance "who the winners are going to be."

He concluded by predicting that IMEC would continue to be a successful example of private-public partnership, based on opportunity seeking as well as risk taking and risk sharing. He emphasized the importance of risk taking: "If there is no risk," he asked, "why would you share it?" He also expressed gratitude for the support of local government, and for the growing number of links with business partners.

Discussion

Dr. Spencer asked how long IMEC's residents stay. Dr. de Proft said visits varied in length, but that he felt scientists get the most out of the experience when they stay at least 1-2 years.

Dr. Wessner asked why, in view of the success of IMEC, it still received government support. The response was that IMEC needed a critical balance of fundamental and applied research, to avoid a 100 percent commercial orientation or to avoid being driven to much closer-to-the-market research—making open innovation more difficult—and that the government was the primary source of support for fundamental research.

Is IMEC Subsidizing Foreign Firms?

Professor Flamm asked what he termed an "impolite question" about the presence in Flanders of large multinational semiconductor companies. Was not IMEC essentially subsidizing research for these firms, none of whom had production facilities in Flanders? Dr. de Proft called the questions "astute and pertinent," and noted that the grants were not discounts on commercial research contracts, but were meant to support fundamental research as a basis for further research programs with industry and with a view on long-term spillover effects for the region. When you dig deeper, he went on, you see payback for the region at many levels. He emphasized the presence of the residents, about 300 bright minds from around the world, spending creative years here, and building up networks. They are all people likely to move up in their organizations, where they will be in positions to make decisions about where to put their R&D centers or other activities. Furthermore, over 200 PhDs are performing their doctoral research at IMEC. IMEC also is interacting with local industry and has furthermore created over 25 spin-off companies, among which are some very fast growers (e.g., Photovoltech). IMEC's activities are also generating strong economic derived impact at the region (e.g., >€42 million of subcontracting to the local industry). The overall impact, being calculated by an external expert company in 2005, is a multiple of the government funding. "Our government is smart enough to understand not to look for direct matches, but to promote some formative behaviors without trying to steer the economy."

ECONOMIC IMPACTS OF SEMATECH ON INNOVATION IN SEMICONDUCTORS

Kenneth Flamm
University of Texas at Austin

Professor Flamm said he would discuss research that seeks to understand the past of SEMATECH, looking primarily at the 1990s, which he called "an important and dynamic period in the semiconductor world," a time when there was increasing global dispersion of technology and production facilities. SEMATECH, which he called a new U.S. experiment in R&D strategy, was put in place in the late 1980s "but really drove forward in 1990s." He said that around the mid-1990s there was a significant acceleration in semiconductor technology, when there was also a global spread of knowledge and expertise in making semiconductors. He said he would focus on microprocessors, because they were the product for which the rate of technological improvement was the fastest in the 1990s, when they were also the dominant single IT product manufactured in the United States.

In 2004, almost half (46 percent) the U.S. integrated circuit (IC) shipments by value made in the United States were microprocessors, compared with 29 percent in 1995, and 37 percent in 2002. For DRAMs, the portion made in the United States was 14 percent in 1995 and 11 percent in 2004. Microprocessors had the highest rate of technological innovation in the 1990s, and the largest input to a PC by value. They also had a big impact on the technical improvement of computers and productivity in downstream IT-using industries. Finally, economists had a very rich data set on microprocessor units (MPUs) which allowed them to perform high-quality research.

A Trend of Price Improvement

He pointed out a notable improvement in the prices of MPUs for three periods: 1991-1995, 1995-1999, and 1991-1999. He used calculations based on common economic price index methodology and price performance improvement in different categories of semiconductor products. In each category, rates of price performance improvements (compound annual growth rates) were significantly greater for 1995-1999 than they were for 1991-1995. These data were convincing because they held not only for MPUs but were consistent "pretty much across the board. This suggests that an underlying factor was at work."

A significant part of this increased rate of decline in prices seemed to coincide with other events, he said. The first was a new U.S. R&D strategy, including formation of SEMATECH in the late 1980s. In 1992, SEMATECH sharpened its focus on manufacturing, especially to accelerate introduction of new technology nodes, using lithography as the benchmark for state of the technology. The goal was to reduce the time between new nodes from 3 years to 2, and its apparent success inspired imitation in Japan and elsewhere.

The Role of SEMATECH

An interesting feature of SEMATECH that is seldom emphasized, he said, is its coordination function. In the early 1990s, the United States developed a National Semiconductor Technology Roadmap. Begun under the aegis of the National Advisory Committee on Semiconductors at a workshop held in 1992, it evolved into a broader attempt to coordinate a complex process of technology development to a point where products would all come online when needed to advance manufacturing. The First National Technology Roadmap came out in 1994, with much of the technical leadership provided by SEMATECH. It was updated in 1997 and has been codified at 2-year intervals since.

In the late 1990s, the roadmap became international and was called the International Technology Roadmap for Semiconductors (ITRS). This change recognized that semiconductor firms were now spread around the globe, and that coordination among suppliers and users had helped to accelerate innovation in the industry. The consensus is that the roadmap has helped maintain the 2-year nodes, he said, and although many people think that 2 years is not long enough to fully realize potential profits on companies' investments, they have not been able to lengthen the cycle because of competitive pressures.

This degree of R&D coordination, he suggested, was a unique structure of great interest to economists. It is the kind of activity that might invite antitrust pressure, he said, but a federal law passed in the 1980s granted limited antitrust immunity for registered consortia like SEMATECH.

The international SEMATECH began in 1995 as a partnership to work on 300mm wafer technology, encouraged by the federal government. This was followed by the recovery and stabilization of the U.S. semiconductor industry. The U.S. government subsidy ended in 1997, and today the share of world semiconductor output accounted for by SEMATECH members exceeds the share when it was formed in late 1980s.

In September 2004, the "international" designation, too, was dropped, though the organization still has many full international members; the most recent to join is Samsung. It has spun off a subset of R&D activities into the International Semiconductor Manufacturing Initiative (ISMI), which walled off access to the "highest tech" activities (e.g., lithography). The main SEMATECH organization has nine "full" members (AMD, Freescale, Hewlett-Packard, IBM, Infineon, Intel, Phillips, Samsung, and Texas Instruments), who also have membership in ISMI. It also has three ISMI-only members who do not get access to full SEMATECH information: TSMC, Panasonic/Matsushita Electric, and Spansion. The first Japanese member was Renesas, followed by NEC in 2006.

But even as SEMATECH went international, the U.S. semiconductor industry's global share of R&D declined as U.S. firms moved more functions offshore. In the 1990s, there was resurgence of semiconductor leadership in U.S. after some years of decline, and U.S. semiconductor firms again moved to the top. Then R&D coordination through the roadmap in the 1990s brought coordination with suppliers in areas where the "best of breed," he said, were no longer located in the United States. After the millennium, increasing offshore competence again led to some increase in offshoring in R&D by U.S. firms.

New Models of R&D Coordination

Since then, he said, interesting new models of R&D coordination have emerged—largely because semiconductors now require very large investments in R&D. One model, subsidized by and located in New York State, is a hub-and-spoke system. IBM, the hub, works with three core partners in developing manufacturing process technology: Samsung, Infineon, and Chartered. Toshiba and Sony are also involved, as is AMD. The model is probably somewhat less open than IMEC or SEMATECH because the partnership is negotiated one on one with other core members. Another partnership is Crolles II, formed by Phillips, STM, and Freescale. This group also has government support and international composition.

What are the benefits of such models? he asked. An obvious one is some acceleration in the rate of manufacturing innovation, such as the new 2-year technology nodes begun in the 1990s.

Benefits of Shorter Times Between Nodes

Another benefit, related to the use of roadmaps, is improvements in price performance, which may be viewed in two ways. One is by engineering efficiencies in products already made—lower price for a given quality or functionality. A second is new capabilities that become possible because of pooled technology. These benefits, he said, are not independent. By shrinking the features of a chip, a company can not only produce the same chip in a smaller area, thereby saving cost, but also the chip can be faster. So while shrinking the technology nodes from 3 years to 2 years gained about 50 percent in price performance, this gain had two components. Roughly half the decline was due to improvement in processor quality, but acceleration in technology nodes also led to acceleration in processor speed. This is because a byproduct of smaller feature sizes is shorter distances between features, which allows for faster chips. Design innovation is needed to make use of greater switching speeds, which is a big factor in user evaluation of processor quality. So this gain in speed, he concluded, was another benefit of the acceleration in nodes, beyond merely reducing manufacturing cost.

The Importance of Manufacturing Gains

Price performance improvement, however, had slowed in the last year and a half as MPUs hit a "brick wall" related to power and heat dissipation in 2004-2005, and this decline coincided with a slowdown in the rate of processor speed increases. A much bigger share of price performance was now due to improvements in manufacturing costs.

He concluded by saying that the R&D coordination that began with SEMATECH and continued through the years of the international roadmap appeared to have created significant economic benefits over the last decade. More recently, the gains made in the manufacturing process have become more important because the rate of improvement in other key components has slowed.

IMEC AND SEMATECH: AN INDUSTRIAL PARTNER PERSPECTIVE

Allen Bowling
Texas Instruments

Dr. Bowling, who manages Texas Instruments' external research activities in silicon technology development, said that in the 2000s his company had "moved from an era of microelectronics to nanoelectronics," routinely producing gates as small as 40 to 50nm. He said that he would talk about where this would lead in the future as the trend toward nanoelectronics continued.

More Dependence on Consortia and Outside Knowledge

At the fabrication facilities of Texas Instruments' Dallas headquarters, the company had adopted the roadmap with 2 years between nodes described by Professor Flamm. Because it takes at least 4 years to fully develop each node, said Dr. Bowling, the company has two to three of them in co-development at any time. In-house technology development programs start about 3 years before manufacturing, and the development group is involved until about a year after manufacturing begins, which amounts to a 4-year period. Altogether it takes about 7 to 12 years to move a new material or device into production. This means that its engineers are more dependent on the long-range knowledge resources of SEMATECH, IMEC, and universities to keep up with current fundamental and applied research and work at the current accelerated pace. They also depend on close collaboration with equipment suppliers.

Texas Instruments has other ways of staying current with technology developments. Beginning about 5-20 years before product qualification, Texas Instruments collaborates more than it ever has with universities, especially through the SRC consortium. Texas Instruments provides direct funding for some short-term SRC needs and is a member of the three SRC consortia. Altogether, these consortia are supported by about $55 million per year from industry and about $20 million per year from the federal government, and other public funding. The program provides funding for about 1,000 graduate students; Texas Instruments supports student preparation generally, on the premise that students represent the future employees for the industry.

The company also promotes partnerships with states to support research infrastructure. He stressed the importance of local government involvement in supporting facilities like IMEC, including a location and a fab structure. State support allows the companies in consortia to focus on running programs, which is the strength of industry. The reason the IBM-centered consortium is viable in New York State is that the state is funding the infrastructure, as does Flanders for the IMEC facility.

Why Texas Instruments Belongs to Both IMEC and SEMATECH

Texas Instruments has been a charter core member of SEMATECH since it began in 1987; there are currently eight core members. It has also been a core member of IMEC since 2004, after following activities in selected programs since 1993. Texas Instruments is willing to pay its membership dues in both SEMATECH and IMEC because, said Dr. Bowling, "it earns a high return on its investment." Both programs spend over $100 million per year on the pre-competitive research needs of the next one to two nodes, and Texas Instruments is able to leverage the knowledge it gains from association with the eight or nine other members.

They belong to both consortia because each has unique capabilities. SEMATECH is best at driving the international roadmap. Members gather to discuss the issues "so the right attention goes to the key gaps for the future." It is also a true industry consortium run by members who each have one equal vote to determine exactly what kind of instrument to work on. The voting can be a rallying point when there is agreement, or a problem when member companies do not agree.

Texas Instruments also belongs to IMEC, where it finds several advantages:

- **Advanced equipment.** IMEC has advanced immersion 193nm and EUV lithography, through close alliance with ASML, which is based in the Netherlands. Their equipment is "above that of any other consortium in the world."
- **Development collaboration.** IMEC has provided much more development collaboration for equipment suppliers. They can tailor individual relationships with suppliers, agreeing to work on co-development and keeping the results reasonably confidential until the capability is proven. After that, they begin sharing it with the other member companies.
- **Focus on fundamental science.** IMEC focuses more on the fundamentals of why things happen. Many research staff are also university faculty, and there are many grad students working at IMEC with strength in advanced device concepts demonstration and testing.
- **Public support.** Public support from Flanders, and smaller amounts from the EU, allow IMEC to keep its research infrastructure at the state of the art, which is critical for the competitive position of global companies.

He also listed several disadvantages of the IMEC system:

- **Less focus on manufacturability issues.** While IMEC's strength lies in advanced device concepts demonstration, it focuses less on manufacturing.
- **Less management control.** Even though it has core members, management reserves right to determine key areas to work on. This means that not all of a member's key issues will be addressed. But IMEC is good at focusing on key issues when diverse opinions exist.

He summarized the advantages of consortia with a sobering statistic. A typical integrated circuit manufacturing flow involves more than 1,000 process steps. The complexity of this flow provides numerous opportunities for collaboration on many challenges, and provides strong rationales for the roles of both SEMATECH and IMEC. Texas Instruments receives unique value from its memberships in both IMEC and SMT, he said, because they each focus on areas of particular strength.

Session IV

Innovation Through Knowledge Diffusion

Moderator:
Mark B. Myers
Xerox Corporation (retired)

LEUVEN AS A HOTSPOT FOR REGIONAL INNOVATION

Koenraad Debackere
K.U.Leuven

Professor Debackere began by noting the rapid increase in scientific collaboration, especially between industry and academia, and he proposed to review research data developed with colleagues about the effects of these links at K.U.Leuven.

Generating Innovation Opportunities

He said that when universities such as Leuven are regarded as knowledge institutes, they can generate innovation opportunities in many ways. These include:

- Startup technology-oriented enterprises formed by researchers out of the science base generated at the research institute;
- Collaborative research with companies;
- Contract research that is based on consulting by scientists who are commissioned by industry;

- Participation by graduate students in temporary practical studies at firms or in jointly supervised thesis projects;
- Advanced academic training of industry employees;
- Personnel exchange, or mobility, between companies and research institutes; and
- Informal links between industry, government, and universities.

He said that interaction was the key word. At one level, interaction takes place at the level of science, such as co-publication between different EC countries. Publication is becoming more important in Europe, as it has been in the United States for many years, and is starting to influence academic and inter-academic behavior in the European community. Universities are increasingly asked for accountability to show how they used public funding and how they can contribute to innovation.

Europe is also seeing the influence the Bayh-Dole Act had in the United States and its effect on IP management at universities. "Some think we're going too far in that direction," he said, "but nevertheless the fact that universities are being held accountable to exploit the IP they generate has given rise to more patents."

He noted an increase in industry funding for university R&D. Germany leads in this activity, at 11.3 percent, with Belgium at 10.9 percent, the EU-15 at 6.9 percent, and the United States at 6.3 percent. Some institutions do much more, he said, which is always a weakness in using general figures, but this trend already influences how university budgets are created and planned. He said that K.U.Leuven had been working hard for 10-15 years to increase its interactions with industry.

Growing Economic Role for the University

The economic role of the university is growing more significant for the region, he said. Its mission statement includes the following goal: "To promote and support knowledge and technology transfer to industry."

This mission is carried out at three levels. At the top are researchers on the payroll. As of 2005, K.U.Leuven supported 974 researchers, a number that had doubled in 5 years. Many of these do research for industry. At the middle level, the university is actively involved in three areas: contract research, spin-off formation and regional development, and IPR and licensing. The third level is industry itself.

Traditionally, he said, the university had two basic missions: research and teaching, which are still fundamental. But in that traditional academic environment, faculty research was done in "almost a pure ivory-tower setting." Nowadays, however, universities in many European countries are charged by the government to create structures and activities that support the commercialization of their

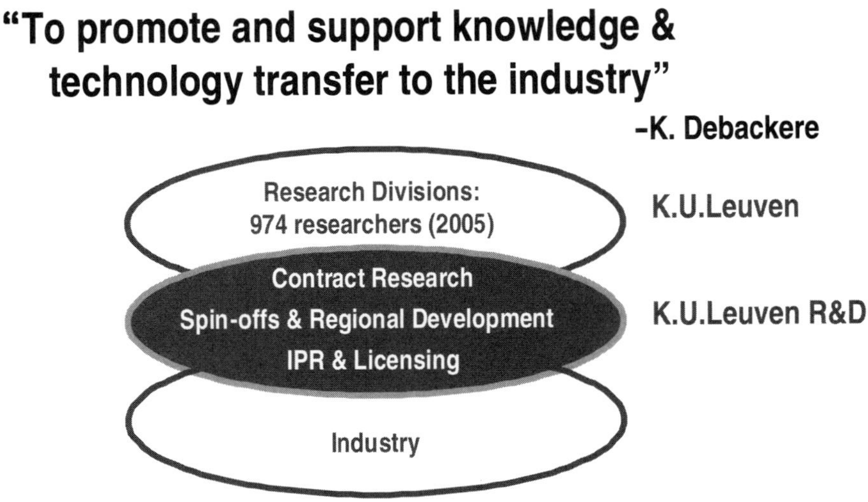

FIGURE 7 Leuven R&D mission and organization.

research. In the Netherlands, he said, universities have a holding company as a separate legal entity to commercialize R&D. This signals a strong strategic intent to exploit research through innovation, but is still somewhat decoupled from what happens within the university.

A Matrix Structure for Leuven

At Leuven, which Professor Debackere considered to be unusual, there is a "full matrix-like structure" that gives academic researchers incentives to collaborate with industry. The academic subjects are divided into three groups: biomedical research, the other exact sciences, and the arts and humanities. Within each are the faculty members and the different departments, "the normal hierarchy where people are recruited and promoted on the basis of their teaching and research abilities."

At the same time, the university has a horizontal structure with about 50 research divisions under the umbrella of a central office of Leuven R&D. The divisions are organized on an interdepartmental basis, and professors of research become members of one of those divisions, under which they can organize their industrial involvement. Any proceeds from their work remain within the division. What drives them, said Professor Debackere, is a desire to be part of a strong research environment where they can compete and collaborate with the best of their colleagues. In order to do that, the university lets them reinvest the income in infrastructure, equipment, and postdoctoral students—in building a strong research environment in the university. "Although this has been criticized as 'social welfare'," he said, "we regard it as the best kind of social welfare, because everything is reinvested in the research." In order to support the divisions and their activities—which include applied research, technology transfer, and the generation of new companies—about 40 people are employed to provide management support, IT support, and consulting on the incubation of new companies.

Advantages of Working with Industry

The literature, he said, has traditionally warned academics against working for industry, on the grounds that it may reduce the quality of their work or pull their interest toward financial gain. Professor Debackere described a study that examined the research groups at K.U.Leuven and other studies that gathered comparable data. The study evaluated academic researchers who collaborated with industry in terms of their publications in ISI-SCIE journals, for both basic and applied research. It also examined the industry involvement and output of control groups of researchers, with similar disciplines and age structures over varying time periods. Consistently, he said, the quality of the academic researchers' work seemed to benefit from collaboration with industry. Groups heavily involved with industry published *more*, not less, basic science work. "This for us is a signal that the interaction between industrial R&D and academic R&D can be a reinforcing one," he said. "Industrial R&D feeds academic R&D with serious problems. So we don't find a perverting effect of academic science."

After mentioning the challenges facing Europe in strengthening the innovation system, he said that research institutes have a critical role to play, which is to transform the university into a regional economic hotspot. He said that in Leuven, more than 100 spin-off companies had already been created. This has led to the creation of "Leuven Inc.," a 600-member network organization where entrepreneurs and researchers do meet on a regular basis. Part of its success in expanding entrepreneurship, he said, grew out of the formation of effective networks. Some of these were horizontal: contact between universities, IMEC, startups, and other "innovation actors." Others were vertical: technology clusters, such as DSP Valley, which focused on design of hardware and software technology for digital signal processing, and L-SEC (Leuven Security Excellence Consortium), an international non-profit network dedicated to promoting the use of e-security.

He summarized the reasons for Leuven's success at commercializing R&D as follows:

- A critical mass of high-quality, internationally competitive research. "This is why IMEC and K.U.Leuven are very strict in their performance assessments."
- An integrated approach to technology transfer, such as incentives for multidisciplinary teams and high value-added services;

- Clear incentives and policies to encourage individuals, research groups, and departments to pursue spin-off opportunities;
- Creation and acceptance of an entrepreneurial climate in a university context;
- A Flemish legal context that is positive with respect to the exploitation of academic research and IP.

"Based on my own experience here," he concluded, "it's the integrated approach that makes it all work."

Discussion

A questioner expressed admiration for the efforts on behalf of the Leuven region. But he expressed doubt about the value spending 3 percent of GDP on research, and asked how that would help create durable and stable jobs for people. He said that the business climate in Belgium in general was not favorable for new firm formation, and asked whether people trying to create spin-offs were expected to give up their jobs to do so. Professor Debackere agreed that the business climate in many European countries, including Belgium, will need to reform in the coming years, and that this would require persistent effort. He also tried to clarify the process of forming a spin-off, which is to create a new employment opportunity for the founder and for those employed by the new firm. Leuven wanted to help people do this, but those moving to a new firm were making a career choice. They could not both do a spin-off and remain in their former traditional job.

AN INDUSTRY PERSPECTIVE: THE CASE OF THE CHEMICAL INDUSTRY

Erwin Annys
Federation of the Belgian Chemical Industries and Life Sciences (formerly Fedichem, now Essencia)

Dr. Annys began by emphasizing the importance of the chemical industry for Flanders and its economy. With only 1.3 percent of the European population (EU25), Belgium produces 8 percent of the EU's chemical products, of which 70 percent is in Flanders. Recalculating the chemical activity per capita brings Belgium into second place in the world. The first place, he said, is for Ireland, which had passed Belgium only 2 or 3 years earlier, "largely due to some differences in economic constitution and possibilities created by the Irish government." He said that Belgium was especially strong in the pharmaceutical sector, with about 40 percent of all pharmaceutical products found in one of the laboratories there.

The Goal of Industrial Biotech

The purpose of his talk, however, was to propose a substantial shift in the way the chemical industry will function. Much of the current industry depends on petroleum-based products, as do the transportation, energy, and other dominant industries. The world is running out of petroleum, and it is time to reduce the world's dependence on it. The objective of Fedichem, he said, is to prepare our industry to be ready in time to combine biotechnology with agriculture, to create and produce new agro-chemical building blocks to provide replacements for the ubiquitous products of petroleum chemistry. This would, in essence, create a huge new industry of industrial biotechnology.

To achieve this ambitious goal, he has begun working with colleagues to create a regional version of SusChem, the European Technology Platform for Sustainable Chemistry. SusChem is formed by CEFIC, the European Federation of the Chemical Industry, and EuropaBio, the European Federation of the Biotechnology Industry, to advance certain goals of the EU, including a more dynamic knowledge-based society (Lisbon, 2000), sustainable development (Goteborg, 2001), and increases in R&D expenditure to 3 percent of GDP (Barcelona, 2002).

European Objections to New Bioproducts

He said at the outset that his vision would be impeded by environmental and philosophical objections to the creation of new chemical and biological products. This, he warned, was based on widespread misperceptions that "could be a major obstacle for the continuous evolution of Europe." He cited studies concluding that European competitiveness was at risk unless more social acceptance is obtained that innovation is "a key driver for future competitiveness." Chemical innovation, in particular, he said, has an "enormous impact downstream," and he asserted that "the only way to grow further in all industrial sectors is by paying even more attention to chemical innovation."

He said that the three pillars of SusChem are industrial biotechnology, materials technology, and reaction and process design. For the first pillar of industrial biotech, the main goals are to create "bio-renewables" to reduce carbon dioxide emissions and conserve fossil fuels. One of the most promising ways to do this, he said, is to use organisms to convert cellulose or lignin to various substances, but the opposition in Europe to GMOs is a "major topic we have to tackle if we want to succeed." He discussed the many bioprocesses and bioconversions that will be necessary, including biopolymers, biopharmaceutics, enzymes, and biofuels. "How can we turn an economy which is now petroleum based into a bio-based economy?" he asked. "By research on the conversion of starches, sugars, and other renewable resources into the same materials derived now from petroleum or into new substances."

For the second pillar of materials technology, he said that SusChem aims to provide marketing, technology guidance, and innovative products. He said that work in this field would strengthen European competitiveness and improve the well-being of citizens. He estimated the growth rate for nanomaterials and nanotechnology at 10-15 percent per year.

As part of his vision, he discussed his "dream" of a house that not only uses less energy, but even generates energy. It would make use of many new materials, such as advanced photovoltaics, self-cleaning facade paints, lighting by white organic light-emitting diodes (OLEDs), and nanofoam insulation. It would use new techniques of energy conversion and generation. "This is the kind of innovation so far rarely presented publicly by the chemical industry," he said.

He listed other areas whose development could lead Europe toward a bio-based future:

- Personalize health care: Here he foresaw new materials for implants, smart drug delivery systems, novel therapeutics, health protection, instant diagnostics, and disease detection sensors.
- Reaction and process design: This area focuses on fundamental enabling technologies, integrating the complementary approaches of chemical synthesis and process design. It contributes all the way from individual reactions to the viability of production plants, and drives sustainable development of the EU chemical and biotech industries.
- New nanotechnology approaches: These include materials with new optical properties, hardness and toughness, and electromagnetic properties; new chemical processes like chemical reactivity and catalytic yield; and new bioapplications through self-organization, reparability, adaptability, and recognition.

In conclusion, he said that his goal was to bring a renewed vision to chemistry research and development in Europe. Without chemistry, he said, the EU could not reach its goal of sustainable development, and this could only be accomplished when all parties and sectors work together.

Discussion

A questioner asked how soon he thought Europe would be ready for a change to a bio-based economy, given the obstacles he had listed. Dr. Annys said that change would take time. He emphasized the need to maintain the existing chemical industry during the changeover, so that the long lifetime of existing chemical installations, which were not designed for 10 years, does not allow a direct change. And a lot of research and scale-up trials are still necessary. "So realistically we are talking about at least 30 years before we will see drastic changes, is my personal impression," he said.

INNOVATION THROUGH KNOWLEDGE DIFFUSION

Paul Ducheyne
University of Pennsylvania

Professor Ducheyne said that his work, which focuses, among other subjects, on eradicating bone infections, depended not only on the diffusion of antibiotics away from the implanted surface but also on the diffusion of knowledge into the environment. This required work, he said, with the sources of innovation, the creation of knowledge, and the culture of universities. "We apply the laws of physics and chemistry," he said, "and also knowledge of biology for the benefit of society. Our teaching gains in depth by first-hand insight into the organizations that apply this knowledge. We also have to be role models in education. Guidance by educators toward implementation of concepts makes for powerful examples in engineering education."

A Broader Training Environment for the PhD

He repeated the observation by previous speakers that research-based training is central to the mission of U.S. research universities. PhD work is very focused, and typically does not span the whole spectrum from fundamental to applied work. But when the lab widens this spectrum to include technology transfer, PhD trainees, too, are exposed to the broad nature of engineering science and its natural collaborations with industry. Academia cannot do the whole job in health science, for example, because valid analyses of interaction between materials and living tissues require larger sample populations than are found in university labs. "The goal is to improve clinical outcomes," he said, "so we need clinical studies, and academia is not well organized to do that."

He listed many different ways to widen the spectrum of education, including industry-sponsored research in academia and off-campus collaboration on production, regulatory issues, and legal matters. Likewise, he said, there are many different routes to a startup. One model is creation of a fundamentally new technology for which an appropriate industrial organization does not exist. The startup helps advance technology, create employment, and build society's wealth.

Diffusion of Knowledge Through Patents

Patents are another essential element in knowledge diffusion, providing the means of transferring knowledge out of academia. The hypothesis, he said, is that top patent holders in the life sciences positively influence the research productivity of colleagues and trainees around them and provide a return on the public's investment in biomedical research.

Once a patent is secured, he said, technology transfer can begin. In U.S. academia, this is assisted by centers for technology transfer (CTT), which help scout for potentially patentable work and manage the application, prosecution, and licensing aspects of patenting. Through the CTT, incentives are created, and the net return is distributed to inventors (about 30 percent) and the university, including individual labs, schools, or research foundations.

He cited the opinion of the European Research Council (2005): "In research, Europe has too long adhered to a defunct model of research utility. It must recognize that the transition to a globally competitive, innovation-driven economy necessarily depends upon the stimulation of fundamental research and its link to the innovation process."

Obstacles to Technology Transfer

Technology transfer is seldom simple, he said, especially for universities without long experience in the process. Academic cultures vary enormously, and the process of technology transfer may encounter various obstacles:

- University leadership may be unfamiliar with tech transfer.
- The CTT does not have expertise in all areas.
- There may be unrealistic expectations of licensing terms or other outcomes.
- Some faculty are not interested, regardless of incentives.

In the biomedical field, programs exist to help bridge the critical gap between fundamental work and product development. These programs allow for proof of principle. Some state-sponsored programs are designed to address this transfer time, such as Ben Franklin Technology Partners in Pennsylvania, which includes company involvement and refundable loans after achieving revenues or 8 years of operation. Bio Advance is another model, primarily for pharmaceutical and biotech, which takes an equity position, with ownership determined on the basis of typical investment paradigms for the industry. At the national level, there are the SBIR (Small Business Innovation Research) grants and the ATP (Advanced Technology Program) grants that facilitate bridging this gap.

In licensing, some universities have requirements, such as continued dissemination of knowledge after the license is granted. The university also cannot be held responsible for the quality of the product. Finally, for young faculty the tenure process must be a priority over commercial activities.

An "Arm's Length Relationship" Between University and Industry

He illustrated the complexities that can arise in academic-industry partnerships, using Orthovita, a biomaterials company, as an example. The company produces a broad materials technology platform that is useful in devising bone substitutes and bone grafts. Although there was a partnership between the University of Pennsylvania and Orthovita, academia and industry have different objectives. This gives raise to "relational cliffhangers." Specifically, the company cannot be the funding vehicle of the university lab, and the university lab cannot be the extension of corporate R&D. "An arm's-length relationship is essential for both parties," said Professor Ducheyne.

He concluded with some principles of the academic mission. First, for academic institutions, it is fundamental that knowledge creation and dissemination share top priority. This implies that knowledge creation will benefit from the quality of education delivered to students at all levels, be it undergraduate, graduate, or postgraduate. Second, a technology transfer relationship between centers of higher learning and their corporate offspring must not be open-ended but confined in time.

Discussion

Professor Ducheyne was asked how he would improve innovation in European universities. He emphasized the importance of the environment in each case, and said, by way of example, that in Holland at least, the "cliffhangers" he mentioned were often inadequately considered. "It is important that universities do not become applied science centers," he said. "Excellence cannot be compromised."

Meeting the Early-stage Finance Challenge

Moderator:
Luc Soete
University of Maastricht, Netherlands
and UN Univ-MERIT

THE TEXAS EMERGING TECHNOLOGY FUND

Pike Powers
Fulbright & Jaworski LLP

Mr. Powers introduced the case of the Emerging Technology Fund in Texas "as an example of what can be done to build innovation." Innovation, he said, is more than activities of technology or business; it is also the process by which it occurs. With that in mind, he and others began in 2002 to create a Texas Technology Initiative designed to take some "conscious objective steps" toward a vision of innovation and capital formation for technology. They persuaded Texas Governor Rick Perry to include in his state-of-the-state speech a request for an "enterprise fund"; the request was successful and funded at a level of $295 million in 2003. A large portion of the early funding went to "save SEMATECH" for Texas in Austin, but subsequent funding created a $200 million Emerging Technology Fund in 2005, which also benefited SEMATECH and other entrepreneurial ventures. Both funds were reauthorized and refunded in the 2007 session of the Texas legislature.

Collaborative University Partnerships

He also described a consortium of 12 U.S. universities (from the Big-12 athletics conference), including Texas, that have formed a virtual Center for Economic Development, Innovation, and Commercialization. The objective is to jointly create tech transfer, research, and share equipment and other resources. "It's the paradigm of the future," he said, "having universities collaborate." The initial pattern for the Center was an athletic conference whose schools would "compete on the football field on Saturday, but collaborate the other six days of the week. The model is designed to break down barriers, structurally change relationships, and enhance communications.

Activities of the ETF

The Emerging Technology Fund with its small size and flexibility has proved to be an effective vehicle for commercialization. While the governor, lieutenant governor, and the speaker of the House approve the work of the 17-member advisory committee that evaluates potential candidates, there are no fund managers and only a small staff within the governor's office to run the program. Each participating region operates a Regional Center for Innovation and Commercialization (RCIC).

The ETF has three sections. $100 million goes to Commercialization Investments, for which the ETF has already reviewed more than 331 applications. Of these, 22 projects at 7 RCICs received $27.5 million.

The second activity is the $50 million Research Matching Grants, intended for industry consortia or single companies who already have some funding from the federal government or other sources. Of 53 requests, 14 projects were approved for $22 million in funding.

Third are the Research Superiority Grants, a fund of $50 million to "discover the best researchers we can find in the country or the world and bring them to Texas." This fund has brought researchers to Texas Tech University, the University of Texas Health Science Center in Houston, and other laboratories.

Mr. Powers concluded with the news that Governor Perry would announce the following week that $10 million of the Research Superiority Fund would go to outstanding researchers in nanotechnology. He then introduced Dr. Goodall, with whom he works on the Texas Technology Initiative.

OVERVIEW OF TXAN: A NEW MODEL FOR RESEARCH COLLABORATION

Randal K. Goodall
SEMATECH

TxAN, said Dr. Goodall, is the Texas Alliance for Nanoelectronics, which is currently being assembled as an alliance with the federal government, beginning

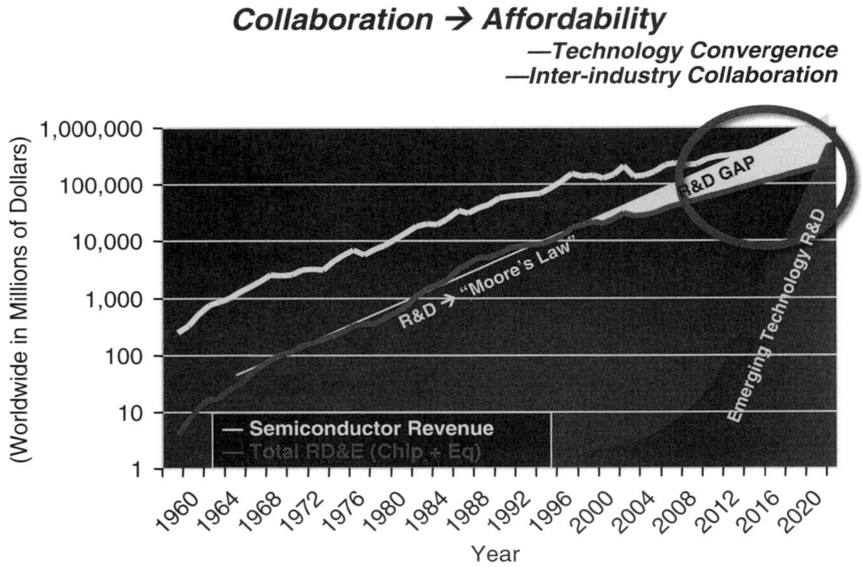

FIGURE 8 Advanced technology R&D challenge.

with the National Institute of Standards and Technology (NIST).[21] He described TxAN as a statewide partnership for building innovative, networked nanotech capability "that leverages world-class researchers and R&D infrastructure and drives regional commercialization of technology." It was the culmination of more than 4 years of work by the Texas Technology Initiative, and builds on regional university and industry strengths. He said it was both a technology-based platform for economic development and an economic development platform for federal and industry partnership.

Key elements included the participation of Texas, which is now the 10th or 11th largest economy in the world, and offered considerable support in the form of the Emerging Technology Fund (ETF) and other large initiatives described by Mr. Powers. This support, he said, "is going to bring to this federal-industrial collaboration more Moore's Law than we deserve. These industries know how to collaborate, and the Texas alliance is all about that."

Collaboration Lowers Costs of Research

Goodall showed an illustration of the future of the IT research over the coming years, showing how collaboration among firms eventually lowers producer

[21]TxAN has subsequently broadened its scope to nanotechnology.

R&D costs. In the United States, the Nanoelectronics Research Initiative (NRI) is located on the time scale far in the future, and the Focus Center Research Program is somewhat closer to the present. The Semiconductor Research Corporation (SRC) was closer still, and SEMATECH was next to "the border between competitive and pre-competitive research." One of the features of SEMATECH, he said, is that it is a flexible organization, so that the manufacturing side of the program could actually be split into a separate industry consortium where the 450mm wafer initiative will occur.[22]

A New Training Paradigm

TxAN was designed to create a new innovation and commercialization paradigm. Within the competitive semiconductor world, it would include a nanoelectronics workforce development initiative that would benefit all participants. This $4 million program would support 160 interns in three categories—those with 2-year associate degrees, those with a 4-year bachelor's degree, and graduate students—to work in the fab under a standardized training routine. The objective was to demonstrate the feasibility of large-scale internships at advanced technology sites, which has not been done before. TxAN also includes the $3 million Nanoelectronics Research Initiative center, described by Mr. Scalise. This third NRI center will provide university research on innovative devices, to be funded initially by the Emerging Technology Fund. Finally, the TxAN Infrastructure Network will operationally link SEMATECH and the TxAN fab to Texas university labs to create a collaborative ~$500 million equivalent "State Lab." Once these programs are in place they will assume the vital task of helping NIST and other units of the federal government fulfill the aspects of their missions that require commercialization. This is an area where the federal government needs the assistance of industry and the states.

Advantages of a "Mezzanine" R&D Center

The Nanofabrication Infrastructure Network will link most of the major university labs in Texas, many of which are world-class. The key, he said, is to link specialty research labs to the Advanced Technology Development Facility (ATDF), a higher-level industry-driven "mezzanine" R&D center—a middle floor between university labs and true industrial manufacturing fabs. The ATDF will specialize in driving the transition of university work to industry. The ~$200 million facility/capability has been developed to support leading-edge CMOS research, including full process control and a full staff around the clock. The target is to upgrade the facility and its processes for the R&D needs of the

[22]The International SEMATECH Manufacturing Initiative has formed and includes the 450mm initiative in its broad equipment and factory productivity mission.

future using both university and private capital funds. There will be state/federal/industry-funded "tokens" to cover processing costs for university researchers. Work that would otherwise be prohibitively expensive, he said, can be done for reasonable cost in this collaborative environment. Universities can take their own laboratory work to an additional level in the fab environment of the industrial real world. This enhanced capability and accelerated commercialization will drive startup formation and corporate-sponsored R&D. They can then tie it together across the state, using web-based documents for all resources, sharing equipment in novel ways, and flexibly using interchange protocols and other "insert" techniques for multiple wafer sizes and other substrates.[23]

The ATDF is now a separate subsidiary company that used to be part of SEMATECH. Because it is separate, each ATDF participant, including companies that are not SEMATECH members, offers support for the core, but has an investment-based capacity allotment and can have private tools and an area for program operations. It can create its own flexible environment and work with suppliers on an individual contractual basis. TxAN, then, is an umbrella of universities, federal engagements, biomedical and energy consortia, and others. It also partners with SEMATECH and Albany Nanotech, where most of SEMATECH's lithography R&D is done.

The Ability to Test Products in a Real Fab

The mezzanine model is the key to modern research, said Dr. Goodall, for several reasons. Manufacturing technology has become critical to the broader field of nanotechnology research, as compared with early nanoelectronics.[24] Having a strong manufacturing focus and environment allows researchers to test in a "real" fab how new techniques will actually work and actually scale. The industry is making a quintillion transistors a year, and is projected to make 60 quintillion a year in 2016. New nanotech capabilities have to be scalable to the level where they intersect an industry. If a new solution is likely to require a billion or a trillion of some technology, researchers need a real facility to find out if they can actually make 100-1,000. The mezzanine concept is designed to make this possible.

Dr. Goodall said that TxAN had received positive responses to its concept from several federal groups, including NIST itself and NASA, and that the TxAN virtual network provided an "overall umbrella on how to stay up with the rest of the world" through directed R&D. The central idea, he said, was to have the federal agencies link some of their mission work with this network. In the end, he said, the concept unites three forces. The first is a multi-hundred-million-dollar state fund

[23]The discussion on how to specifically initiate the State Lab continues with key stakeholders in Texas, and a definitive approach is anticipated by the end of 2007.

[24]See the earlier presentation by Kenneth Flamm for a discussion of the importance of manufacturing-based research.

that drives driving commercialization. The second is the in-depth federal research of NIST, NASA, or others. These two mutually enable the third, university laboratories networked around a core mezzanine facility—this triad attracts and supports the corporate and startup R&D that drives economic development.

State Money + Federal Depth + Industrial Management

The process currently under discussion, he said, is that the university system in Texas would develop management offices, tap into the research and university lab infrastructure, and tie these functions to the ETF as a mezzanine facility that is fully operating with major industrial players. The ETF and other funds can support pre-commercial work, build up the infrastructure, and recruit the best talent. The university system offers management and governance.

Advisory board processes, too, were under discussion, building on the long history of SEMATECH working with NIST. The concept allows a federal agency to see TxAN as an extension of itself, differing from the custom of developing new knowledge through outside contracts.

A Paradigm for Other States

He summarized by defining nanoelectronics in the broad sense now intended by TxAN. Microelectronics has already introduced a wide range of miniaturized specialty materials, including many applications for biotechnology. He said that TxAN was addressing the technologies that would emerge over the next 20 years throughout this new realm of the very small, from nano-electromechanical systems to "bio-nano." The participation of NIST at the outset is central, he said, because a strong focus on measurement and standardization is crucial to sustain commercialization. "Texas has all the components it takes to put this together," he concluded, "and we're quite confident we can make it go. Beyond this, we believe that this is a replicable paradigm that can be used by other states."

Discussion

A questioner asked why TxAN was bringing the university and industry together. Dr. Goodall answered that the idea was not to eliminate university labs or teaching but to provide an industry-like middle step between university labs and actual industrial use. He gave the example of the ATDF nanofabrication facility, where several leading bio-nano device makers need silicon-based technology to build filters and other devices. They could build them in a four-inch university lab, but if they want to try for an FDA-approvable process, or build prototypes, or begin mass fabrication to test the reliability of the process with high volume, they need the capabilities offered by ATDF. At the same time, the work at ATDF is still directed by university-based researchers. Mr. Powers commented that it was an advantage to have industry driving the policy. Government officials would be most likely to support a plan in which the business community plays a dominant role, with university support. "We believe in getting everybody at the table," said Mr. Powers, "so the elected officials can see that everybody's in the same boat going in the same direction."

Another questioner asked what metric TxAN would use in Texas to show progress. Mr. Powers said that their current report was the initial stage, and they were asking the legislators that same question in regard to the ETF: How do you want us to show what we've done with the taxpayers' money? Straight job creation was not the best measure, though it is important, because new approaches always lead to disruptions and lost jobs as new jobs are being created. It was challenging to measure all those processes in a way that the non-technical citizen can understand and accept.

FROM UNIVERSITY RESEARCH TO UNIVERSITY SPIN-OFF: EXPERIENCES OF VUB

Bruno de Vuyst
Vesalius College, Vrije Universiteit Brussel (VUB)
and Lawfort Brussels

Professor de Vuyst represented VUB, a Flemish university in Brussels that places emphasis on creating spin-offs and operates on a "project cycle basis." That is, they start by supporting fundamental research and then examine the outputs of that research as intellectual property for potential commercialization, identifying and studying possible applications. "It is only the last 5 percent of the work that can be valorized, that may be spun off," he said. "A spin-off may then lead to new questions for fundamental research that will feed back into the cycle."

He said that his university was the last in Flanders to start a spin-off fund, but that this could be considered an advantage, since they had been able to learn from the experience of K.U.Leuven and University of Gent. After only 30 years of existence, he said, this university of about 10,000 students had been able to develop an effective research program that generated on average seven to ten patents per year.[25]

"Market-oriented PhDs"

In Flanders, he said, the traditional focus on fundamental research is giving way to a new paradigm. The university system has de-emphasized traditional

---

[25]The VUB is an outgrowth of an older, French-language university founded in 1834 (Université Libre de Bruxelles), both dedicated to freedom of inquiry. It became a free-standing Dutch-language institution in 1969 with a motto that "science triumphs over darkness."

PhD programs in favor of "market-oriented" PhDs. "There is no automatic career future for traditional PhDs," said Professor de Vuyst. "We have mobility programs for researchers. We have centers of excellence in engineering, technology, economics, and medicine. Valorization is the 'third function' of universities now. We have to bring our knowledge to the community. That is a big sea change for us, and it required a lot of work."

Larger reforms in the higher educational system are still needed, he said. At present, students can enter any university at age 18 for only about €500 in tuition and choose their degree freely. "We end up with far too many people who are not marketable and hence frustrated because they end up with degrees that the market does not seek. The cost of this is high to taxpayers, and is not sustainable. It is politically not correct to raise the issue, but this is something we will have to deal with at some point, as budgetary constraints will prevail."

Needed Reforms for IPR

He turned to issues of IPR, saying that awareness of their importance is growing. On innovation, the public sector is more or less meeting the Lisbon goals, although the private sector is not. The related topic of patent protection faces issues that are even more complex. Patenting costs, for example, are about four times higher than in the United States. There is still no European Community patent, and there are considerable barriers to pan-European suits against patent infringement. The EU court of justice had recently ruled against that practice, which means that a patent must be defended country by country, discouraging patent development. Nor is there any grace period for publication of scientific results, meaning that university professors who think they have a patentable idea must refrain from publishing in the scientific literature, at a cost to their reputation. Finally, he said, European countries needed to remove software from the TRIPS copyright framework.[26] This was no longer an issue in the United States, he said, and needed to be resolved quickly in Europe.

He concluded by saying that Flanders, although it was doing better than much of Europe in developing an innovation-friendly culture, still had many

---

[26]The WTO's Agreement on Trade-Related Aspects of Intellectual Property Rights (TRIPS) and the Berne Convention placed software protection solely under copyright. Following the U.S. Supreme Court decision in Diamond vs. Diehr, the United States has allowed software protection through patent law; something not provided for in TRIPS or the Berne Convention. Since then, the USPTO has continuously granted patents for software.

Professor de Vuyst points out that in Europe, the same evolution is hindered by the exclusion language in article 52 of the European Patent Convention (EPC). At the time of the conference reported in this volume, there was an expectation that, because of the recommendations of European Patent Office's own expert committees, art. 52 EPC would be changed in the EPC 2000 text and the exclusion language would be abandoned, or at least diminished. This did not happen, as the member states did not follow the recommendations but stayed the text as is, with the result that software "as such" remains unpatentable in Europe.

issues to address. There is some private equity involvement in new high-tech firms, he said, but not nearly enough entrepreneurship or venture capital risk taking. Much more attention should be given to preparing new spin-offs for growth and survival, which now takes much too long. Finally, the biggest issue for Flanders is the shortage of management that is qualified and experienced enough to lead the growing number of new firms into the future.

Discussion

A questioner asked whether the high cost of patenting in Europe can be reduced. Professor de Vuyst said that this was impeded by articles of EU treaties with respect to state aid, but that they would have to find a way to bring the cost down, perhaps beginning with some form of support for translation or development of claims and claim writing.

COMMERCIALIZING UNIVERSITY RESEARCH: THE ROLE OF THE U.S. SBIR PROGRAM

Charles W. Wessner
U.S. National Research Council

Dr. Wessner began by describing the "innovation imperative"—the modern realization that innovation is a prerequisite for maintaining a competitive position in the global economy. A key in responding to this imperative, he said, is the knowledge that small businesses anchor the innovation process by adopting, developing, and commercializing new ideas to a disproportionate degree. Small size and flexibility are ongoing themes in any discussion of innovation, as are the roles of the university and the faculty researchers involved in knowledge dissemination. He urged the symposium participants to think about "issues of scale," and the power of the small but dynamic firms that bring innovations to the marketplace.

The Accelerating Pace of Global Competition

He addressed the issue of global competition in general, and the rapid increase in scale and effectiveness in China, where becoming the dominant global manufacturing center is a national goal. India brings its own scale advantages and an especially entrepreneurial high-value culture. France and Finland are renewing and funding technology programs, with France committing to a new €1 billion innovation agency. The key point, he said, is that the pace of competition is accelerating. To keep up, nations need to strengthen their science and technology base, maintain an open system of trade, create incentives for R&D and knowledge transfer, and foster innovative small businesses.

Small Business as Key Driver of Innovation

Small businesses are a key driver of the U.S. knowledge-based economy, he said, having generated 60-80 percent of net new jobs over the past decade. In 2003, the most recent year for which full data were available, small firms created 1,990,326 new net jobs, while large firms shed 994,667 net jobs. Small firms also employ 41 percent of all high-tech workers, including scientists, engineers, and computer workers, and produce 14 times as many patents per employee as large patenting firms.

The U.S. economy is large enough so that small firms have unlimited potential to grow, as illustrated by the cases of Intel, Microsoft, and Google. This is seldom true for small economies; in Sweden, for example, no new large firms have developed since 1970. But even well-run small firms with promising ideas face major challenges to growth. Potential financial backers, such as venture capital firms, have the problem of understanding and forecasting what a nascent firm can do. Firms propose many good ideas, and even more bad ideas. For the private equity community, the problem is one of sorting and discovery, and success is never guaranteed.

The Danger of the "Valley of Death"

The greatest danger to a small firm comes at the stage of development after the end of public support and before the availability of private support, which is traditionally provided by venture capital (VC) firms. This stage, often called the "valley of death," is critical for the small firm that must develop a prototype, develop a commercializable product, and organize a sound management team. (See Figure 9.)

With the recent shrinkage of seed funding for new firms and the shift of military budgets away from basic research and toward weapons testing, however, this "valley" may have widened in recent years, making the route to commercialization even more daunting. Once an inventor's personal funds and the support of friends or "angels" is exhausted, there remain significant and expensive tasks before a VC firm is likely to be interested in participating.[27]

"We have a powerful myth in the United States that free markets can do it all," said Dr. Wessner. "And to many, suggesting a role for government in this process is confused with 'picking winners'. In a sense, we are always picking winners with our public policies, and what we want to do is find the best young companies and help them perform."

[27]For more discussion of this trend, see the discussion by Professor Good earlier in this volume.

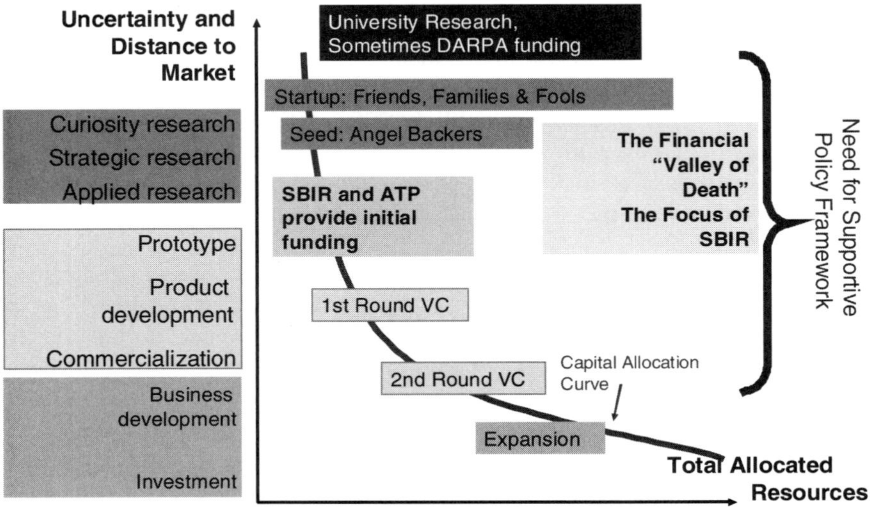

FIGURE 9 U.S. innovation curve.

An Environment Friendly to Entrepreneurs

Dr. Wessner noted that the United States does at least maintain a policy environment that is friendly to entrepreneurs, and that the culture as a whole does support innovation. And the United States has been effective in promoting innovation through a variety of mechanisms, such as industry-led consortia for standards and joint research, university-based research, joint ventures with the Advanced Technology program (ATP), and the Small Business Innovation Research (SBIR) program, which grants support for proof of principle and prototyping. Although industry provides the greatest share of funding for early-stage technology development in the United States, the federal government is estimated to provide between 20 and 25 percent.

By contrast, he said, some features of European culture have restrained innovation, such as punitive bankruptcy laws and a culture of caution that avoids the risks inherent in small-firm formation.

Crossing the Valley with the Help of SBIR

The United States does have a role in helping firms across that valley of death, Dr. Wessner said, and the most assistive mechanism is the SBIR program. This program, created in 1982 and renewed in 1992 and 2001, has a budget of \$2.2 billion, and participation by all federal agencies with an annual extramural R&D budget greater than \$100 million is mandatory. The SBIR is not a procure-

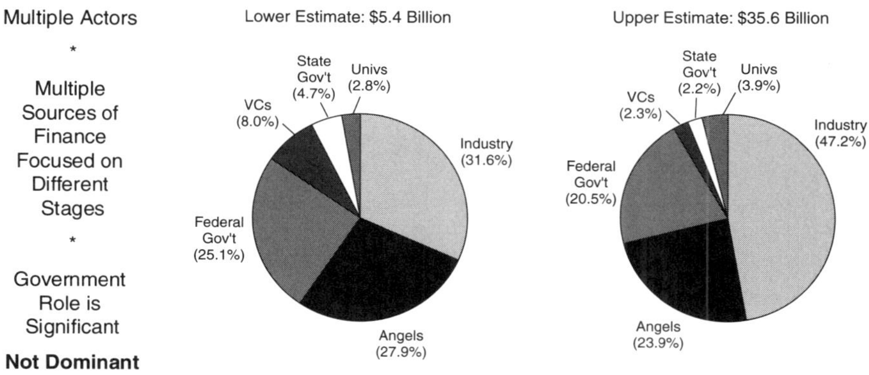

FIGURE 10 Funding sources for early-stage technology development in the United States.
NOTE: Figures based on 1998 data.
SOURCE: Lewis Branscomb and Philip Auerswald, *Between Invention and Innovation An Analysis of Funding for Early-Stage Technology Development*, Gaithersburg, MD: National Institute of Standards and Technology, 2002.

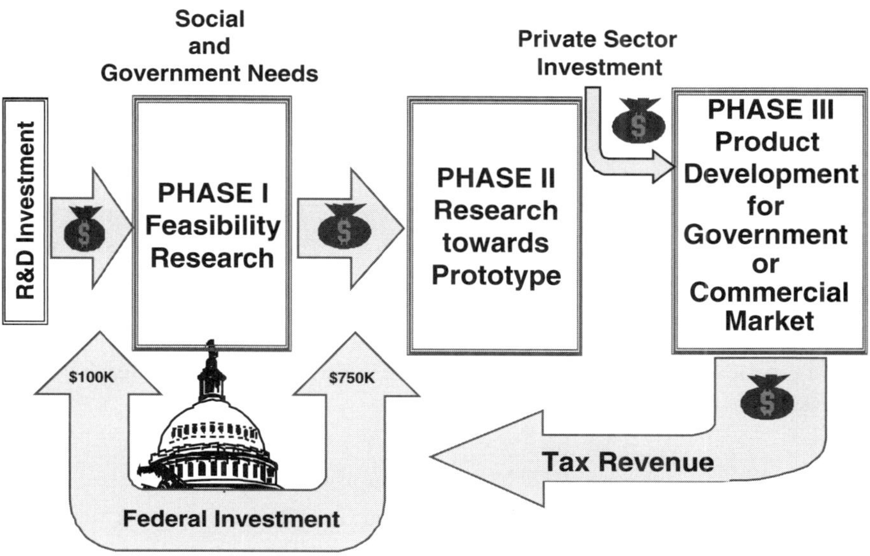

FIGURE 11 SBIR model.

ment program. Instead, each agency sets aside 2.5 percent of that budget for awards to small business.

About 12-15 percent of applicants receive a $100,000 award, primarily for feasibility research. If their work is successful, they are eligible for a second award of $750,000, typically for development of a prototype or equivalent output. In the third phase, no money is awarded, but two outcomes are common: someone may buy the company, or an investor may provide funding.

There are several advantages to the SBIR program, he said, that might serve European systems as well. Because of the set-aside mechanism, agencies do not have to ask for money each year. Another advantage is a "crowding-in" effect. Economists have predicted a crowding-out effect when government involves itself with the private sector, but firms that have won SBIR grants tend to attract *more* private-sector funding, not less.

Benefits for Government, Industry, and Academia

There are other benefits for both governments and entrepreneurs:

- For governments, SBIR can be a flexible tool for solving agencies' problems. Agencies receive proposals that are industry-initiated, so that ideas flow from the bottom up. Program ownership rests with many agencies, not a single "tech agency," and the SBIR may change incentives within organizations for those who wish to change.
- For entrepreneurs, SBIR provides additional research funds of $850,000+. It asks no repayment, royalties, or dilution of ownership, but grants recipients rights to IP developed using SBIR funds and helps attract additional capital through a "certification effect."

Dr. Wessner noted that SBIR also helps universities in several ways, first by linking them with industry. Universities opposed this at first, but in the mid-1990s, following implementation of the Bayh-Dole Act, they began to embrace the concept and create their own technology transfer offices. Also, SBIR awards make it easier for faculty researchers to create new firms without giving up their university posts. Universities help diversify and grow the job base for all types of employment, and participation in SBIR helps validate research funding and generates returns to society.

Could an SBIR Model be Helpful in Europe?

SBIR might be a useful model for Flanders to consider, suggested Dr. Wessner, given they ways in which it links key elements of the national innovation system, increases employment, and helps commercialize new knowledge. It would require no new funds, while encouraging innovation and entrepreneurship and providing a signal of quality to other potential investors.

Some European observers also feel that an SBIR program might be productive for Europe. As Jens Rostrup-Nielsen has written, "We need to do more to encourage the creation of new high-tech firms. Small companies need to be more than just subcontractors to their larger counterparts in FP6. . . . SBIR works in the United States; if it is not possible to establish such a system here, then perhaps that is one of the reasons why we are not doing so well economically."[28] Dr. Wessner noted that SBIR "experiments" have already begun. Finland has adopted a Phase I program, following consultation with the National Academy of Sciences, and Sweden has initiated SBIR-type mechanisms. The UK has a program similar to SBIR in concept, the Netherlands has a pilot program under way, and Taiwan has begun an SBIR program.

Resistance to SBIR Features

On the other hand, some European officials, said Dr. Wessner, profess approval for SBIR while hesitating to adopt the features that make it effective:

- Real competition: Phase I and II competitions create clear signals about the quality and commercial viability of an innovation; some officials resist a mechanism that yields a low approval rate.
- Stable funding: A percentage set-aside creates a stable funding stream which, over time, creates a large portfolio of projects with higher potential for success. Some officials prefer budget-line funding, which is often subject to political uncertainty, small scale, and consequent low returns.
- Applicant ownership: Tying programs directly to agency missions can strongly motivate entrepreneurs. European technology-related programs are often run by a national technology agency, walling them off from the innovation process.

"SBIR is a proven mechanism in an uncertain game," Dr. Wessner concluded. "It preserves a market orientation, connects small businesses and new ideas to national needs, and improves market operation by bringing more competition. The market orientation is essential in commercializing ideas that really work and bring value to society."

Discussion

A questioner asked how one can know that the SBIR competition process really selects the best companies, and whether there is a guarantee that they will succeed. Dr. Wessner acknowledged that there can never be a guarantee for

[28] In European Commission, Strengthening Industrial Performance, Luxembourg: Office for Official Publications of the European Communities, 2003.

companies or ideas in a free market. "But sometimes a dry hole is valuable," he said, "because it demonstrates that something doesn't work. Occasionally there is a huge payback, and just a few of these pay for the whole program many times over."

FUNDING FLEMISH INNOVATION: GOALS, MECHANISMS, AND RESULTS

Rudy Aernoudt
Department of Economy, Science, and Innovation

Mr. Aernoudt discussed the range of mechanisms available in Flanders for funding small firms, from the seed and early startup stages to the expansion stage when a firm is ready to enter the market. He said that many venture capital firms avoided risky investments in the smaller businesses, either because they wanted a better guarantee of return on their investment or because they did not want to invest in a scheme in which the government had any role. So the Department of Economy, Science, and Innovation looked at the U.S. SBIR program and found that 70 percent of all seed and early-stage investments were supported in some way by the government, and decided to design some analogous mechanism to provide government backing for innovative firms.

A Government Program for a Market that Does Not Work

In surveying investment patterns in Flanders, Mr. Aernoudt found that the average ROI on seed money was only 2.3 percent, and that 85-90 percent of all money invested in private companies was going to later-stage and employee buyout firms. "The market alone did not work," he said. "We had to convince ourselves that if the market did not work, the government would not necessarily do better. What we wanted was to give enough incentives to make the market work."

Flanders set up a range of instruments to match the size of firms. For the smallest firms needing only seed money, Flanders set up the "Win-win loan" mechanism. This offers small loans up to a maximum amount of €50,000 per SME or entrepreneur. Investor risk is practically nil, with the borrower paying back the loan through yearly tax credits of 2.5 percent, up to a maximum amount of €1250 per year. In the event of bankruptcy and liquidation of the firm, 30 percent of the loan will be paid back by the government. The interest rate is set between 50 percent and 100 percent of the legal interest rate, and the maximum duration of the loan is 8 years.

For slightly more advanced firms in the later seed and startup stages, the government created the Flanders Innovation Fund (VINNOF), which is a purely public fund. It is funded by the sale of shares and is budget-neutral for the government. IMEC receives some of its financing from VINNOF. A problem, said Mr. Aernoudt, is that it is very small, with only about €5 million distributed so far among 20 firms.

To finance early-stage and expanding firms of larger size, the government created the Arkimedes program that is modeled on the U.S. Small Business Investment Corporation. "Flemish people have money," he said, "but don't like risk. So we fund this program by offering bonds with a 100 percent government guarantee." Investors may also invest cash in the fund, receiving a 90 percent government guarantee plus an 8.75 percent tax credit per year. If Arkimedes should lose all its money, investors are still guaranteed a positive ROI of 2 percent. The bonds have been sold and the proceeds divided into four Arkimedes funds, or ARKIVs.

Mr. Aernoudt said that the Flemish mechanisms of debt financing had a theoretical basis in the Modigliani-Miller Theorem, which states:

- The value of a firm is determined by its earning power and the risk of its underlying assets, and is independent of the way it chooses to finance its investments or distribute dividends.
- Under certain assumptions, it makes no difference whether a firm finances itself with debt or equity.

As a consequence, there is a bias toward debt financing that leads to undercapitalization.

He concluded that Flanders' programs were able to make good use of abundant public money by using these co-investment schemes, and that they brought fiscal advantages to those who invested in innovative SMEs. His summary of the situation was that "Flanders is a paradise for innovative companies."

Discussion

A questioner asked Mr. Aernoudt how his department would measure success. He responded that success would not be measured by tax revenues, because "we have perverse taxes." He said they were using such milestones as the number of innovative companies supported, and that the problem of good research results not being converted into real companies had been solved by using available money. They also measured the absorption rate, by looking at how well the money was used; evaluated the effectiveness of investment behavior by analyzing the default ratio of the companies; and measured innovation capacity by examining the innovative actions of companies.

YOUNG TECHNOLOGY-BASED FIRMS IN BELGIUM: THE IMPACT OF POLICY INSTRUMENTS

Bart Clarysse
Vlerick Leuven Gent Management School
Ghent University

Professor Clarysse described his efforts to quantify the impact of public policy on young technology-based firms (YTBFs) in Belgium and to answer the question, why do some firms grow faster than others? He began by saying that he and a colleague[29] had identified YTBFs by looking for firms that:

- Were academic spin-offs
- Were in the portfolio of VC funds
- Had received R&D grants
- Were part of a random sample of all startups in high-tech and medium-tech sectors.

They identified and studied 235 examples of small firms in operation between 1991 and 2002, doing a retrospective reconstruction of the early growth path. They were looking for factors that distinguished fast-growing firms from those growing slowly or not at all. They found that the definition of high-growth firms varied with the growth measure used, because the picture is so complex. They focused primarily on the measures of revenue growth, employment growth, and asset growth.

How to Account for Rapid Growth

The first task was to try to account for rapid growth. They looked at a variety of potential determinants, including founding conditions ("imprinting"); initial resources, including people and their experience; and market strategy, including focused vs. diversified and local vs. international.

They were surprised to find that for firms in general, venture capital funding by itself had a negative impact on growth. Small amounts of venture capital were "significantly negatively associated with growth in revenues and employees." However, firms able to attract more than €1 million within 18 months grew fastest. So Professor Clarysse concluded that there were two categories of firms: those that were able to produce a business plan that attracted a lot of capital, and those that had to "bootstrap" themselves with very little start-up capital. Firms winning large amounts of VC usually had business plans based on micro-electronics or similar fields, with clear exit strategies.

[29]This report was based on work by Professor Clarysse and Mirjam Knockaert.

Rapid growth, he said, was not equal to sustainable growth. In fact, the faster-growing firms had the highest failure rates. Growth was sustainable only for firms that created "recurrent/stable" revenue streams before burning all their cash, and several of the high-employment-growth firms failed. A combination of investor acceptance and market acceptance, with a focus on the bottom line, most often led to sustainable growth.

The Experience of Firm Founders

The second task was to examine the experience of firm founders, concluding that firms grew faster when their founding teams were experienced. Teams of two to three founders with a mix of commercial and technical experience were the most successful—a finding that he said has "more explanatory power than the VC variable." Only about 16 percent of the firms had this mix.

Commercial experience of the founding team was important for early growth—so important that he did an in-depth analysis of founders. The solo entrepreneur was generally not successful. The "kinship team"—brothers, mother-in-law, etc.—was often a team of convenience or comfort and usually did not endure or find capital. The most successful teams were "organic": the team was already in place before the business opportunity arose; members might have met at school or previous jobs. The fourth type was the "matched team," with members of similar qualities who came together specifically for the venture. These did well for the short term, but were not sustainable. Team members came into personal conflicts and split up again.

The Key Factor: International Experience

The third factor they examined to explain growth was the international experience of the company. They found that firms with international experience from the outset were associated with high early growth in revenues and total assets, but that there was no significant effect on employment growth in the first years. They tried to compare these observations with existing theories, including organizational learning theory, and collected data on all stakeholders and key partners. They then fit these data with control variables, such as ambition of the entrepreneur, resources at founding, and industry sector.

After extensive data analysis, Professor Clarysse concluded that experienced, multidisciplinary teams with international exposure were necessary for growth. While most policy measures had focused on the need for equity financing to bridge the gap between startup and commercialization, he felt it was useful to look at small-firm growth from a human resources point of view as well.

He concluded by calling for a "rejuvenated industry policy" that recognized the importance of internationalization in determining the success of young technology-based firms. More specifically, he said, this internationalization is most significant when present from the early stage, and when it is accompanied by founders from an international background and the participation of international partners.

Discussion

A questioner asked why he had measured success in terms of revenue, rather than profits, and whether a business that went public early was more or less successful than one that went public later. Professor Clarysse said that they had chosen revenue as a measure because it was more likely to generate employment than profits, which could be tied to specific products rather than steady growth. As for the timing of an IPO, he said that only 6 of their sample of 235 firms went public, so that the measure would not have been helpful.

CONCEPT AND EVALUATION OF THE ADVANCED TECHNOLOGY PROGRAM

Marc Stanley
National Institute of Standards and Technology

Mr. Stanley, director of the Advanced Technology Program (ATP), began by describing its mission, which is "to accelerate the development of innovative technologies for broad national benefit through partnerships with the private sector." He, like several other speakers, emphasized the seriousness of the U.S. seed funding gap and the "dwindling high-risk investments" in new firms, which includes decreased spending on research by industry. These conditions gave value to the question of whether the federal government should be involved in the national innovation system.

The first condition he emphasized was that "the large U.S. venture capital market was not focused on early-stage firms." Only 1.65 percent of total VC investments of $20.9 billion in 2005 went to seed or early-stage projects.[30]

He illustrated the hypothetical progress of a young firm that typically faces its most severe cash flow pressures just when sources of capital were most scarce—when crossing the "valley of death." Because this valley coincides with the stage just after technology creation, when a technology is being developed into something useful, it represents the period when a good idea may find a use in society and a commercially valuable form.

[30]National Venture Capital Association, 2005.

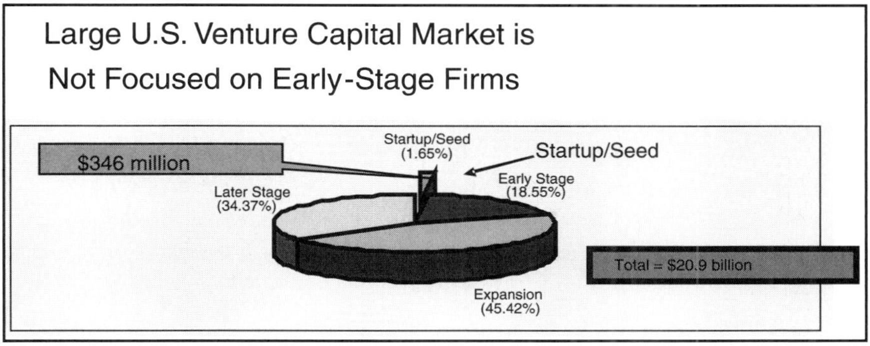

FIGURE 12 Breakdown of U.S. venture capital by stage of development.
SOURCE: Based on 2005 data from National Venture Capital Association.

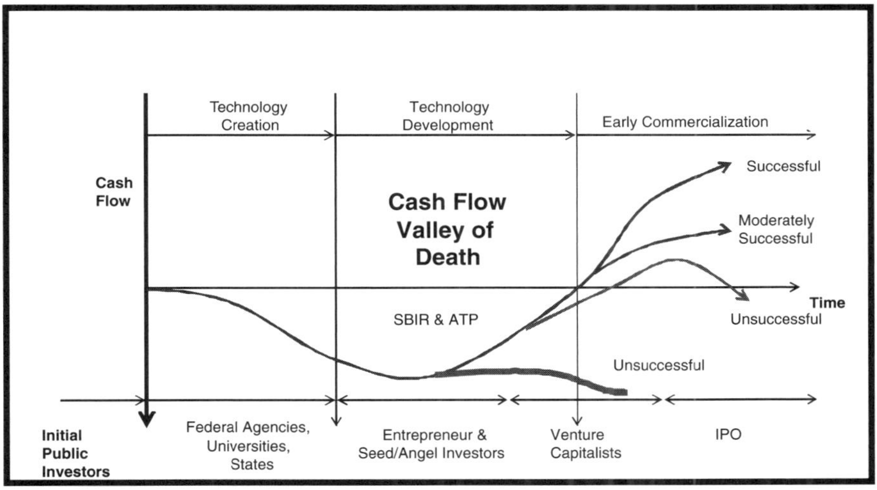

FIGURE 13 Public-private funding transition between innovation and invention.
SOURCE: Adapted from L. M. Murphy & P. L. Edwards, *Bridging the Valley of Death: Transitioning from Public to Private Sector Financing*, Golden CO: National Renewable Energy Laboratory, May 2003.

The Rationale for the ATP

These conditions, he said, provided the rationale for the ATP, which is designed to help young firms through that transition. Some of the key features of the ATP are the following:

- Industry leads in proposing and implementing projects.
- Project selection is based on not only technical but also economic merit.
- Projects must have well-defined goals and sunset provision.
- Project selection is competitive, based on peer review (fewer than 15 percent are chosen).
- The ATP is part of the National Institute of Standards and Technology (NIST), a major federal agency responsible for national measurement standards.

Mr. Stanley emphasized that a good ATP candidate not only has a sound technical premise, but also a business plan, even in its early stage.

Since the formation of the ATP in 1990, said Mr. Stanley, it had compiled a positive record of investing in the category of high-risk young firms. The ATP had made awards to 768 projects, which included 218 joint ventures and 550 single companies. Some 66 percent went to small businesses, and included the participation of more than 170 universities and 30 national laboratories. Of the total award amount of $4,371 million, about half was provided by industry. About 1,400 patents had been created as a result of these awards.

Applicants may file for ATP awards either as single firms or as joint ventures. For single firms, there is a 3-year time limit and a maximum award of $2 million. The company pays indirect costs, and large companies must invest at least 60 percent of total project cost. Many small firms prefer to apply as part of a joint venture with a larger firm, however, which brings advantages: the time limit rises to 5 years, universities are more likely to participate, and there is no limit to the size of the awards, which average $10 million. These ventures may become partnership networks that include universities, consortia, and research labs, as well as the company. They also pay for the expenses of graduate students, and play a key role in the research goals of the federal government and the economic strategies of state governments.

A Confusing Policy Background

In general, he said, public-private partnering in the United States takes place against a mixed and often confusing policy background. On one hand, the Bayh-Dole Act and subsequent amendments have encouraged formation of new companies based on university research, but some universities have policies that restrict technology transfer. More generally, policymakers do not understand how limited is the amount of VC funding available to the smallest firms or the potential importance of these firms to job formation and economic growth.

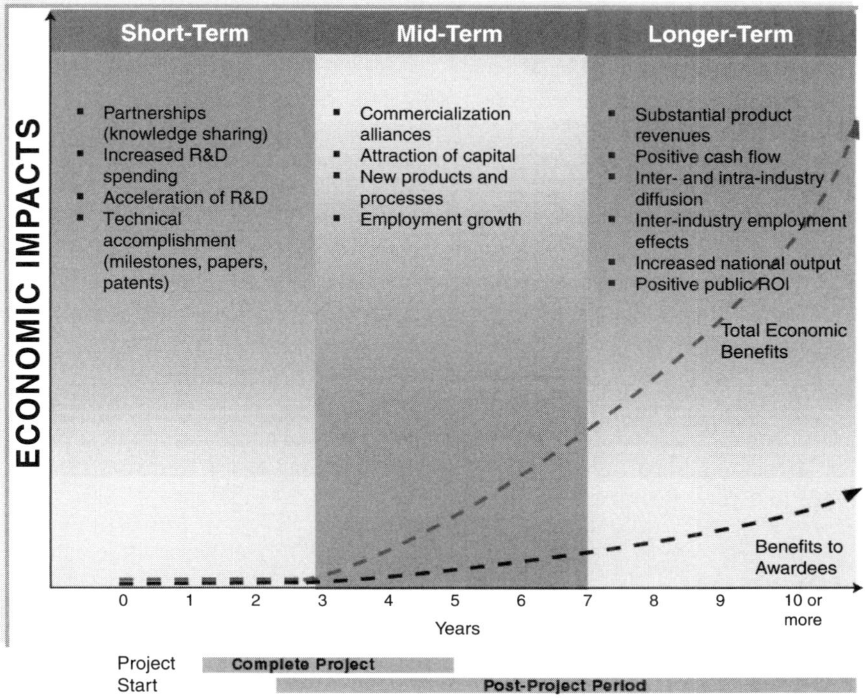

FIGURE 14 Outputs, impacts, and outcomes occur at different times.

Mr. Stanley said that the impact of the ATP, both on award recipients and on the economy in general, had been studied extensively, and was found to be considerable.

The short-term and medium-term benefits for individual firms, for example, included the creation of partnerships, increased R&D spending, the attraction of new capital, and employment growth. Broader benefits included substantial product revenues, increased national output, and positive public ROI that make a difference beyond the benefits to the single firm.

A History of High Returns

Altogether, he said, the program had delivered a net return of almost $1 billion from a government investment of about $2.1 billion. He cited the results of a recent survey of 36 projects which totaled $79.2 million in government investments and delivered total revenues of more than $971.9 million. Revenues and savings resulting from products or processes incorporating technologies developed under ATP awards were said to exceed $2.7 billion.

He emphasized that the ATP spent considerable effort on continual and rigorous self-evaluation, and gave several results of current findings:

- Forty-five percent of projects had brought new technologies into commercialization.
- Ninety-five percent of projects reported that ATP funding allowed for greater risk taking or longer time horizons.
- Eighty-five percent of ATP projects involved collaboration.
- ATP projects had produced more than 1,700 scientific publications and 1,400 patent applications.
- Sixty-six percent of ATP projects were led by small companies.

He concluded by calling the program a success in accelerating the development of technological advances and becoming a "strong causal factor" in moving innovations from the idea stage to the marketplace.

THE CHALLENGE OF COLLECTING GOOD EVALUATION DATA

Bart van Looy
Flemish Policy Research Centre for R&D Statistics (SOOS)

In 2000, the Flemish government decided to create a research center dedicated to the evaluation of its science and innovation policy. This center was charged with developing a system of indicators to quantify the results of R&D at Flemish universities, research institutes, and industry that could be used by policymakers to support appropriate science and innovation policy for Flanders. As a result, the Policy Research Centre for R&D Statistics (Steunpunt O&O Statistieken, or SOOS) was created in January 2002.[31]

SOOS began its activities with studies in three key areas: bibliometrics, technometrics, and innovation. As data sources they used the Web of Science (WoS) and ISI Proceedings databases and various sources of patent and innovation data from EC and OECD surveys. They also developed their own databases and have become "the only place in Europe where all these data sources are found together."

Creating Indicators

In order to do their assessments, said Professor van Looy, SOOS had to develop an appropriate IT infrastructure and create S&T indicators for the gov-

---

[31] In 2007, the Flanders government approved a second generation of policy research centers, resulting in, among other things, SOOS becoming SOOI (Steunpunt O&O Indicatoren).

ernment. They then launched specific studies of science and technology domains and science policy for the Flemish government, as well as original research to measure the effectiveness of policy measures and translate them into incentives. For the government they also began to publish the biannual *Flemish Indicator-book on Science, Technology and Innovation* and provide indicators and discussions of the meaning and dynamics behind the figures. They track such activities as R&D and innovation expenses by the business sector, sales share due to newly introduced products, and distribution of R&D by field. Among the long- and medium-term services for the Flemish government is an additionality analysis that asks, is public funding crowding out private investment? "We want to learn a lot about S&T policy," he said.

"Scientific Services": Are They Good for Universities?

Following a major debate in 1991 over the role of the universities, SOOS is making a significant effort to examine the role of entrepreneurial universities within innovation systems and the impact of legislative frameworks. Since Flanders created a new "triple mission" for its universities, adding "scientific services" to the traditional activities of education and research, SOOS is trying to learn whether this mission is feasible, realistic, and justified. Is there added value for society? What is its overall impact—does it hamper or encourage collaboration between academia and industry? Part of SOOS' objective, said Professor van Looy, is to help the universities ensure that new entrepreneurial activities do not jeopardize the traditional missions. At the same time, if the triple mission continues, he wanted to be sure that universities install procedures that result in a fair return for researchers and research groups.

Some Positive Effects

SOOS has already discovered a positive effect of university involvement in "science services" in the form of increased patenting. This effect was difficult to measure, he said. Many academic patents do not show up in patent statistics because the applicant is the company. Nonetheless, they demonstrated that for the period 1991-2001, when the university acted as assignee, patent activity increased, stimulating technological activity. No negative impact was observed on the collaboration between universities and companies, as measured by the transfer of ownership rights.

He said that understanding innovation was complicated because it involved so many kinds of activities and talents. A contribution of SOOS was to combine many data sources, link activities in diverse fields of S&T, and find patterns based on different forms of information spanning the fields.

Long-term Commitment to Evaluation

SOOS had also examined the relationship between technological performance and human capital, asking the question, Does educational strength (number of PhDs in S&T) contribute to high-technology performance? The answer was yes, there is a distinctive and considerable impact. To arrive at this answer, the researchers had to "disentangle causality," to be sure that the human capital was driving technological performance, rather than the other way around.

Professor van Looy concluded by praising the Flemish government for recognizing the value of evaluating its innovation policy, and for its long-term commitment to the data, infrastructure, and people required. "This work could not be done," he said, "without our proximity to scientific research, alignment with international standards, and ability to network both locally and globally."

Discussion

A questioner asked what kind of jobs could be expected from the work he was describing. Professor van Looy said that "you have to have patience if you want growth—more than 2 or 3 years." For an immediate, direct impact, he said, one could open up a flower shop. But to gain employment benefits from high-technology policies, a commitment must be made over a period of 10-15 years.

He was also asked if the databases he described are publicly available. He said that SOOS had a licensing agreement that did not allow them to make the databases public. For patents, he said, they try to apply open-source logic as much as possible, but they could not infringe on license agreements they signed with researchers 5 years earlier.

Professor Good turned to a more general discussion, commenting on the importance of assessment for any programs. The success rate of the ATP is high, she said, partly because the assessment of candidates is thorough at the point of application.

She also mentioned the importance of entrepreneurial activity to industrial research. Too many companies are interested only in profitability, she said, without having a consensus on how to measure the quality and utility of their R&D. "It all comes down to people," she said. "If you're going to evaluate R&D, you have to assess the talent that's put into the pool." Dr. Spyns observed that Flemish programs are also careful to scrutinize proposals at the front end.

Innovation in the Political Context

Professor de Vuyst commented that an unfortunate aspect of evaluation is that skeptics can claim that researchers get the result they want. Evaluations are useful as support tools, he said, to make the case, but the significant force is the long-term political commitment. This is difficult to gain from politicians who have short-term horizons. In the Flemish case, he said, the support of the politicians had followed a series of historical accidents. The question was how to keep the policies in place. For the Netherlands, he said, election time was drawing near, and the innovation platform had dropped out of the debates, replaced by urgent short-term concerns such as reducing taxes. A goal was to use evaluation tools and other data to make the point that investing in knowledge is as important as investing in other basic structures.

Is There a Long-term Commitment to Innovation?

Professor Flamm said that the sums of money being spent on Flanders' innovation measures were large for the size of economy. He asked if there was political support for the programs to continue. Dr. Spyns noted that the commitment was part of a government agreement to spend an additional €60 million each year, at least until the current government's term ended in 2009, in addition to the investment in the innovation fund (VINNOF) of €150 million over 2 years. He also cited the publicity and awareness campaigns in the media. "The word innovation is everywhere," he said. "We are pushing its importance into the minds of people." Dr. Wessner asked Mrs. Moerman if she agreed. She said that Flanders has both the institutional structure and political consensus, and that she was certain the support was there. "I think that knowledge is linked with identity in Flanders," she said, "in the same way we have a commitment to defense. What we need more of is leadership to fine-tune the policy measures across the system. We haven't seen a lot of that."

Session VI ———

Flemish Strategic Research Centers

The following presentations to the U.S. delegation described the three other strategic research centers of Flanders Inter-University Institute that follow the model of IMEC.

FLANDERS INSTITUTE FOR BIOTECHNOLOGY (VIB)

Lieve Ongena
Flanders Interuniversity Institute for Biotechnology (VIB)

Dr. Ongena, the senior science adviser for the VIB, said that Belgium had a long tradition of excellence in the life sciences. For example, Marc van Montagu and Jeff Schell at the University of Ghent were the first scientists to introduce a foreign gene into a plant. Walter Fiers was first to publish the full DNA sequence of a gene. And Désiré Collen discovered the gene for tPA, which has now found application, through licensing to Genentech, in dissolving blood clots and restoring circulation to stroke victims.

A Determination to Turn Knowledge into Societal Benefits

Despite many milestone discoveries, she said, there had been virtually no translation from the university laboratories to the economic growth of Flanders. This led to the strategic decision to invest not only in knowledge, but also in a mechanism to "turn research results into patents, new companies, and economic growth." In 1995 the government funded VIB as a non-profit research institute charged with this mission. This "institute without walls" brought scientists from Antwerp, Ghent, Brussels, and Leuven under the virtual umbrella of the VIB in a joint venture.

The mission of the VIB is to invest in basic research, to train young researchers, and especially to invest in technology transfer. It is also charged with explaining its mission to the public.

VIB now has 850 scientists and technicians who gather once a year from their respective universities for a conference. It has 250 PhD students, who earn their PhD at their home universities in 4-5 years. Once they graduate they must leave the VIB. "We don't try to keep them," she said. "If they can find money to support their work, then they can come back." The VIB itself does not grant a degree.

Partnerships with Universities

VIB has a 50-50 partnership with the four universities, where its affiliated faculty work in 60 different research groups in nine academic departments. The total research budget is €60 million. Half of that comes directly as a strategic grant from the Flemish government, €6 million from industry, some from the universities' match with the VIB, and some from international sources.

The research portfolio emphasizes molecular biology, cell biology, genetics, microbiology, the 3-D structure of protein, bioinformatics, and systems biology ("a fashionable term for the plant in its environment"). Its researchers also study biomedical areas, primarily cancer, neurobiology, inflammation, cardiovascular research, and Alzheimer's disease.

A Mandate "To Be Excellent"

Their core facilities serve their scientists and the rest of Flanders: a microarray facility, genetic services facility, protein services facility, nanobody services facility, a bioinformatics training and services facility, and a proteomics core facility. Their mandate from the government was "to be excellent," and Dr. Ongena said that they took this seriously. Their scientific output for 2004, as measure by publications, showed an impact factor of 10; when the VIB opened, the impact factor was 5. They began with 700 scientists, and have expanded to 850. When they began, VIB scientists published 16 papers a year in top journals; in the past year they had published 65, a 400 percent increase during a time when the staff grew by only 21 percent.

A Strategy of Licensing

The main emphasis is on the transfer of knowledge into societal benefit. If a discovery is patentable, the VIB usually licenses it to a company. "But if the platform is wide enough, we set up our own company." They had four companies operating, including dVGen, which is developing uses of a microscopic worm, which has genes similar to human genes, for drug discovery; Peakadilly, which does crop design; and Ablynx, which uses a camel antibody as a tool for drug targeting. The licensing revenue promotes VIB's growth and generates money for research. VIB has also given rise to a fifth startup called SoluCel, a small company in Finland, and two days before the workshop it formed a sixth company called ActoGeniX, which uses the Lactococcus bacterium that grows in the human intestine as a living drug delivery tool. GMIV, a government investment company, had invested €11.5 million in the company, which was also supported by a cluster of Dutch and German investors, some VCs, and VIB.

Together, VIB's startups employed more than 280 people, a growth of 100 in 4 years, and had raised more than €220 million in venture capital. When d4Gen went public, it raised €34 million.

The VIB has also invested in a "bio-incubation" center. "If we have an idea," said Dr. Ongena, "we can start a company tomorrow. There is also a biocluster at Ghent Technology Park employing 320 scientists who work for VIB and other companies.

Reaching Out to Society

She said that reaching out to society was an important aspect of the VIB mission. "We want to reach people at all levels. The press and media, of course, and policy makers, but also teachers and students (we need more scientists), doctors and patients." She described a Scientists@work school project that brings groups of 10 to 15 students into one of their labs for a half-day project so "they can see what a career might be, how enthusiastic our scientists are. We need to attract young people. Students want to go into marketing and sales to make money, so this is a major issue for us."

She summarized by saying that Flanders was succeeding in catalyzing the life sciences, thanks to "lots of players": the university, government, angel investors, industry. One indicator of success, she said, was that the VIB had been able to keep costs to about €1 million per record of invention, and €2 million per patent. "That's good," she concluded, "on a worldwide scale."

Discussion

Dr. Ongena was asked if there were interactions between the VIB, IMEC, or other institutes. She said that this was just beginning, as each institute established its own technology. Multidisciplinary collaboration was becoming more and more important, she said. In the case of biotech, for example, the gene can be investigated from many points of view, from patient care to pure physics to microelectronics. With IMEC they planned a collaborative project to grow nerve cells on a chip and measure electrical and chemical changes at synapse.

Professor Good asked how the VIB managed its plant genetics in the face of EU resistance to genetic engineering. Dr. Ongena said it was very difficult, but they were trying. They had held an exhibition called Genes on Your Plate to explain what genetic modification is all about—that when a farmer crosses strains of corn, he is manipulating genes.

FLEMISH INSTITUTE FOR TECHNOLOGICAL RESEARCH (VITO)

Dirk Fransaer
Flemish Institute for Technological Research (VITO)

Dr. Fransaer said that his institute was unusual in the Flemish context in not having links to universities. This is because it was formed in 1988 during the nation-wide reorganization of government when the responsibility for the areas of "energy" and "environment" including the related research activities moved to the jurisdiction of the regions. What is now VITO—including materials research, non-nuclear energy, and the impact of radiation on the environment—moved from the federal government to Flanders. VITO was situated outside the capital region, in Mol, because of the possible danger of exposure to nuclear materials.[32] In 1991 VITO became an autonomous public research company owned by the Flemish government. It is funded by the Department of Economy, Science, and Innovation, but more than 80 percent of its work is done for the Ministry of Environment and Energy.

While the emphasis at IMEC is on cutting-edge research—7 years from the market, he said, the mission of VITO was to work with Flemish industry, and 80-90 percent of all Flemish companies are SMEs employing fewer than 50 people. So its work was practical and solution-oriented, with considerable attention to the day-by-day needs of individual firms in complying with regulations and laws. Of the staff of 510, about 90 have PhDs. The work of VITO is subjected to external review, and their budget for 2006 was about €68 million.

VITO's Activities

VITO's activities span some of the most sensitive fields of R&D in a continent closely concerned with every aspect of the environment:

- Environment: VITO is one of the top three European centers in in-situ *soil analysis and decontamination*, he said, and his job is to help companies deal with contamination and compliance issues. The other areas of emphasis

[32]The Belgian Nuclear Research Centre is located in Mol.

are *waste treatment and reuse* and *air purification.* The Institute is asked to help devise solutions for companies in many compliance issues, as well as to perform research at all levels.

- Energy: One strategy of VITO is to develop better ways to decentralize sources of energy and move away from dependence on petroleum. Among top-priority topics are hybrid vehicles, underground natural gas storage, solar energy, wind energy, and biomass. Belgium has made a commitment to ending its use of nuclear energy by 2014.
- Materials: Among the high-priority research topics for VITO are plasma-technology, laser applications, biomaterials based on ceramics and powder metallurgy, and "rational use of raw materials."
- Remote sensing: The emphasis here is to use remote sensing for environmental monitoring.
- Certification: Flemish policy is to reduce CO_2 emissions by 7.5 percent from 1990 levels, which all companies must follow. VITO was charged with establishing benchmarks for companies and verifying compliance.

INTERDISCIPLINARY INSTITUTE FOR BROADBAND TECHNOLOGY (IBBT)

Wim de Waele
Interdisciplinary Institute for Broadband Technology (IBBT)

The IBBT is the youngest of the four strategic research centers, said Mr. de Waele, founded just two-and-a-half years earlier. The research teams making up the virtual center, however, had been in operation for about 20 years. This center focused on ICT, but had no formal definition or limitation on the meaning of broadband. It embraced not just hardware and infrastructure but also the applications and services that reach into the community.

The Need for a Foundation of Research

Mr. de Waele said that the origin of IBBT lay in the 2001-2002 crash of the dot coms, which affected Flanders as it did the rest of the world. He and his colleagues witnessed two trends: (1) The large companies with research facilities here, including Philips, Siemens, and Alcatel, all scaled down their operations; and (2) some of the local high-growth software companies crashed in spectacular fashion. Because Belgium is a small country, he said, they didn't have the buffer of a large economy, so that when some of those companies went under, the failures had a large psychological impact. In the area of ICT, they saw that it, like biotech and micro-technology, needed a foundation of ongoing research that was both market- and company-focused. "We had missed that," he said.

The IBBT is based in Ghent, but has a virtual structure with research teams in the member universities. To design its programs, the founders analyzed the entire universe of available ICT research, not just the computer aspects but also legal aspects, privacy issues, patent law, communications, user analysis, and related fields.

As a result of this analysis, the institute adopted the mission of developing multidisciplinary human capital, and interdisciplinary, demand-driven basic research in ICT subfields. The primary focus was on ICT, software development, and broadband. They used the institute format to provide a single umbrella for available resources. The focus was on applications of wireless communications, he said, because "that's where the future lies."

The original plan was to distinguish between fundamental and applied research but, Mr. de Waele said, "I found it impossible. All of our work is very user-driven and close to the market and companies."

IBBT started out by selecting 13 groups, with their professors and research teams, which had proven track records in both academic and business areas and experience in working with companies. The institute consisted of 100 PhD staff members and 400 researchers altogether at the universities. They had already published more than 400 papers in A-listed journals.

Carrying Ideas to the Market

Given the mission of carrying ideas to market, the organizers asked themselves how they were going to put that into practice and structure their research. What domains would receive priority: the latest protocol on cable or another level in the value chain in application development? They decided to focus on finding solutions to complex problems that are not easily copied or taken to other places in world. They also wanted to help the economic fabric of the region, so they focused on health because the population is aging and ICT can play an important role in providing services. He saw this as a market-pull approach that fit global society challenges. IBBT also works on issues of mobility, new media, e-government, and enabling technologies, such as next-generation network architecture. One project was to develop a communications platform for interactive home care.

"Interdisciplinary, Strategic Basic Research"

Most of the work was interdisciplinary, strategic basic research. They might build a platform not just as a deliverable in a bilateral contract, but as an output that could be deployed in different countries. They emphasized partnerships in areas of basic research, working in close collaboration with companies. They used a 50:50 funding scheme similar to that of the other interuniversity institutes, with good success in generating joint solutions to complex problems. They had launched about 100 spin-off companies, and had about 30 development projects under way.

He commented that Flanders was an appropriate location for the IBBT and for ICT research in general. "Flanders is fourth or fifth in the world in Internet connectivity," he said. "Everyone has cable." The two largest ICT companies had a "duopoly," he said, with one controlling cable, the other DSL. "They control the market."

The IBBT had been in existence for only 2 years, sustained by a grant of €17 million. Additional revenue came from contract research, and their target was to double the value of the grant in 10 years with their own revenue. Mr. de Waele did not foresee a dependence on the EU. "Their funding is so bureaucratic," he said, "that it is almost more effort to get their funds than it's worth." IBBT did not receive extra subsidies in addition to the yearly government grant or give companies any money.[33] Companies pay half the expenses of a joint program. Clear rules about sharing IP are set out in the contract. "We can use the IP for scientific purposes, they can take it to the market," he said. "If they take it to market, there is some kind of return for us as well so we don't violate EU regulations."

Using Licensing to Recoup Expenses

The institute has two ways to recoup its expenses: joint development programs and income from its own research that leads to spin-offs. There were then four spin-offs in the pipeline. IBBT plans to take a share of a spin-off in the form of licensing revenue. They did not want to create a portfolio of holdings; their goal was to create as many companies as possible, and recoup money only when the spin-off had real revenues. He anticipated taking only a low licensing percentage of 5 percent. "We're basing this on what has been done at universities, and may tweak it as we learn."

Adding Value to Companies and the Flemish Economy

Mr. de Waele said that the primary objectives of IBBT were to add value to companies and to the Flemish economy. They also used academic excellence indicators, but these were secondary to the number of spin-offs. They would be launching a business incubator in the next year, similar to the biotech incubator.

Developing Research Consortia

They also developed research consortia with companies, which so far included 86 organizations in different fields that reflected the fabric of the ICT industry in Flanders. These ranged from Cisco Systems, which employs 1,000 people in

[33]Companies can apply to IWT for subsidies.

Belgium, to a very small startup of 20 people specializing in a new protocol for sensor communication. These partnerships came about through open calls for projects, the first of which was issued just 3 months after IBBT was formed. "We just decided not to waste too much time building a strategic plan, because we had a sense of urgency," said Mr. de Waele. They had so far done four calls, with good responses. Most of the companies responding had already worked with academic research and trusted the idea of collaboration.

The next step was to facilitate the resulting consortia. The consortia were especially useful when multiple people were proposing the same idea from different places. The consortium provided a mechanism for those people to work on the same idea but from different angles. The role of IBBT was to facilitate the dialog between companies and research teams in building the consortia, much like IMEC's strategic programs. "We're somewhat more market driven," he concluded. "We ask the companies what their needs are, and try to pool them."

Discussion

Mr. de Waele was asked whether there were any supercomputers in Flanders to support fields such as bioinformatics. He said there was a grid computing initiative that was multidisciplinary in nature, facilitating not only biotech but also other disciplines. IBBT was also a member of an e-research group studying an investment plan for supercomputing and grid computing.

He was also asked about the participation of foreign companies. He said that any companies were welcome to participate as long as they were active in Flanders. "We're not too strict about that," he said. "Our goal is to stimulate economic activity here, whether foreign or domestic. Borders are a thing of the past in terms of scientific collaboration. We just want to bring the best people together and let them work."

III
RESEARCH PAPER

China's Drive Toward Innovation

Alan Wm. Wolff[1]
Dewey Ballantine LLP

October 4, 2006

On the road into Suzhou, a large rooftop sign proclaimed "Development is an Immutable Truth" in English under massive Chinese characters. The message was from Deng Xiao Ping, and although it has been translated in various other ways over time, the message is unmistakable: There is one acceptable path for China and that is economic development. This sentiment might not seem remarkable in any country seeking to industrialize. But in China, it carried special force.

[1]Author: Alan Wm. Wolff, Managing Partner, Dewey Ballantine LLP, Washington D.C. office, was a member of The National Academies' Board on Science, Technology, and Economic Policy (STEP Board) from 1997-2004. The Council of the National Academy of Sciences and the Governing Board of the National Research Council then designated Mr. Wolff as a lifetime "National Associate" of the National Academies. From April 2005 to the present Mr. Wolff is serving as a member of the Committee on Comparative Innovation Policy: Best Practice in National Technology Programs. This article draws on a presentation made to the "Conference on Regional and National Innovation," co-hosted by the U.S. National Academy of Sciences and the Flemish Government, in Leuven, Flanders, Belgium, on September 21, 2006. The author gratefully acknowledges the splendid research capabilities of the Dewey Ballantine LLP International Trade Group whose research efforts are overseen by partner Thomas R. Howell, its lead research experts Rachel Howe and Maki Hishikawa, and the devoted assistance of Gabriel Key and Maxine Walker. I also benefited from confirmation of my conclusions gained through exposure to discussions with many experts on the subject of Chinese innovation at the Levin Institute in July 2006, when this project was already well formed. The views expressed here are of a legal practitioner, former government negotiator, and an advisor to major companies engaged in trade and not those of a formal scholar, which accounts for the perspectives, insights, and undoubted shortcomings, but also its perspective. The views are solely those of the author, and not necessarily those of his firm, its clients, or the National Academies.

Deng's message has been refined over time by his successors but never diluted. In 1999, Jiang Zemin stressed the importance to China of innovation:[2]

> *In today's world, the core of each country's competitive strength is intellectual innovation, technological innovation and high-tech industrialization.*

The current leader, Hu Jintao, continued this emphasis, incorporating innovation as the guiding direction for national policy.

> *[We should give] priority to independent innovation in S&T [Science and Technology] work, take efforts to enhance S&T innovation capability, increase core competitiveness and [strive to make] S&T innovation with Chinese characteristics a reality.*
>
> *We must aim to be at the forefront of the world's S&T development, speed up the building of [a] national innovation system,...strengthen the coordination of economic policies and S&T policies, [and] create a policy environment beneficial to technological innovation, high-tech development and industrialization.*[3]

In the Chinese system of governance, statements like these shape national and local policies. There is a singularity of purpose in China rarely found in Western governments even when comparable sentiments are on occasion expressed.[4]

---

[2]Jiang Zemin, General Secretary of the Communist Party of China Central Committee, keynote speech, National Technological Innovation Conference, August 23, 1999.

[3]Hu Jintao, General-Secretary of the CPC Central Committee, November 27, 2005, as quoted at *<http://www.most.gov.cn/>*.

[4]Cf. TOKYO (XFN-ASIA)—*Prime Minister Shinzo Abe has said he will make government leaner and foster innovation to pilot the economy through the threats posed by a shrinking and aging population. In his first policy speech to Parliament since taking office this week, Abe said technological advances were vital to ensure sustained economic growth. . . . "It is possible to bring about economic growth even if the population shrinks. Through innovation and open-minded policies, I promise to infuse vitality into the Japanese economy," said Abe.* Japanese PM aims for leaner government, innovation 09.29.2006, 02:39 AM, *Forbes* Magazine Web site.

Cf. also the following White House announcement: *America's economy leads the world because our system of private enterprise rewards innovation. Entrepreneurs, scientists, and skilled workers create and apply the technologies that are changing our world. President Bush believes that government must work to help create a new generation of American innovation and an atmosphere where innovation thrives.*

On April 26, 2004, President Bush announced a series of specific measures to inspire a new generation of American innovation—policies to encourage clean and reliable energy, assure better delivery of health care, and expand access to high-speed Internet in every part of America. By giving our workers the best technology and the best training, we will make sure that the American economy remains the most flexible, advanced, and productive in the world. White House Web site, *<http://www.whitehouse.gov /infocus/ technology/>*.

And for the European Union: *"Ten priority actions to achieve a broad-based innovation strategy for the European Union. The aim of the Communication is to present 'a broad based innovation strategy for Europe that translates investments in knowledge into innovative products and services'. The Com-*

These pronouncements of China's top leaders have been accompanied by an amazing array of detailed policy measures at all levels of government. China is already well into a process of industrializing. What Beijing has decided to do is ". . . to move China from an imitation to an innovative stage of production . . . from 'made in China' to 'made by China.'"[5]

Innovation is seen by China's leadership as essential for China to continue its economic growth, maintain political stability, support advanced military capabilities, and retain China's global trade and geo-political power. In short, for China, innovation was to be a policy of unrivaled importance. The prominence given to innovation as state policy and the resources devoted to it make it unwise to discount the likelihood of success despite the many obstacles still to be overcome. Moreover, it should not be forgotten that throughout most of recorded history Chinese science and technology was superior to that of another nation and even that of all other nations combined.[6] China has not forgotten.

Foresight is rarely granted to us. A world view in 1950 would have been unlikely to have anticipated Japan's achievements in manufacturing and international trade during the next 40 years nor the successful creation of a European common market, and its joint endeavors like the Airbus.

ARCHITECTURE OF THE CHINESE INNOVATION POLICY SYSTEM

This is a fascinating time in Chinese development. With respect to innovation policies and their implementation, there is much emerging. China has released policies encouraging science and technology development in the past,[7]

---

munication proposes a ten-point programme for immediate action to make the business environment more innovation-friendly." MEMO/06/325, Brussels, September 13, 2006. The action plan memo is available on the Europa Web site, *<http://europa.eu/rapid/pressReleasesAction.do?reference=MEMO/06/325&format=HTML&aged=0&language=EN&guiLanguage=fr&&&&&&&&>.*

[5]Kathleen Walsh, Testimony before the U.S.–China Economic & Security Review Commission, April 21, 2005.

[6]For example, China was familiar with cast iron and coal firing in the fifth century B.C., and was probably smelting ore with coke by the thirteenth century A.D. Europe, by contrast, did not cast molten iron before the fourteenth century A.D., and smelting with coke did not come into common usage until the last years of the eighteenth century. Paper was invented in China and the technology was transferred to Europe via the Islamic countries. The Chinese were producing gunpowder in the ninth century A.D. and firearms in the eleventh century, long before Europe. Printing presses with movable type were being used in China in the ninth century A.D. when Europe was still in the Dark Ages and over five hundred years before Johann Gutenburg "invented" the printing press in Germany. Chinese scientific concepts known to have diffused to the West include quantitative cartography, magnetic science, and the double-acting principle of rotary and longitudinal motion. See Joseph Needham, *Science and Civilization in China* (five volumes), Cambridge: Cambridge University Press, 1954-1986; Fernand Braudel, *Capitalism and Material Life 1400-1800,* London: Harper Colophon Books, 1973.

[7]China began issuing science and technology specific programs in earnest in the 1980s. Significant previous efforts include the Key Technologies Research and Development Program (1982), the 863

but China's new Innovation Policy is different, both in depth, because it involves long-term technology planning beyond single-ministry technology development programs, and in breadth, because its execution is broadly dispersed among a half-dozen ministries at the central-government level, as well as at the provincial and local levels, and is strengthened by powerful and ongoing policy tools, such as government procurement, that give the broad policy initiative teeth.[8]

The basic documents for China's Innovation Policy include China's Medium and Long Term Program on Science and Technology Development (2006-2020), the unfolding of the 11th Five-year Plan (2006-2010) and in all probability the new National IPR strategy. Even when underlying documents are not yet published or fully elaborated, the tenor of the strategy can be identified in the statements of central government and provincial officials.

The overarching document to China's innovation planning and strategy is the Medium and Long Term S&T Program. To achieve the objectives of the plan, a variety of policy tools are being used to promote, favor and reward indigenous innovative technologies. An overall goal is to increase R&D spending to 2.5 percent of GDP by 2010, a doubling over the current rate. The 2010 target is comparable to the current rate of spending by the United States, is 0.6 percent less than Japan's and 0.6 percent more than the European Union's. The expected doubling in China's spending will be accompanied by the implementation of key state projects ". . . launched to generate important strategic products. . . ."[9] Their breadth and scale are huge. In the United States, comparisons might reasonably be found in American investments during the period 1945-1991 in telecommunications, space exploration, communication, aeronautics, and energy.

The Medium and Long Term plan identifies 16 key state projects covering a number of priority sectors, and represents an enormously ambitious series of undertakings:

- Core electronic components;
- High-end general chips;
- Basic software;
- The technology for manufacturing extremely large integrated circuits;

Program (1986), the Torch Program (1988), the Spark Program (1986), and the 973 Program (1998). China added strategic emphasis in the 1990s with incentives, offered specifically through the Tenth Five-Year Plan for Science and Technology (2001-2005).

[8]Many of the 99 regulations supporting the State Council's innovation policy (the long-term science and technology plan) will be drafted by NDRC (approximately 29), many others by MOST (approximately 17), some by MOF (approximately 21) and the Ministry of Education (approximately 9), and the remainder by miscellaneous other government agencies, including MII and MOFCOM.

[9]State Council of the People's Republic of China, *Outline of the National Medium-and Long-Term Program on Scientific and Technological Development, (2006-2020),* February 9, 2006, Provisional Translation from Chinese on file, original available at: *<http://www.cas.cn/html/Dir/2006/02/09/13/70/88.htm>*.

- New-generation broadband wireless mobile telecommunications;
- High-end numerically controlled machine tools and basic manufacturing technology;
- Development of large oil and gas fields;
- Large nuclear power plants with advanced pressurized water reactor high-temperature gas-cooled reactors;
- Control and treatment of pollution in water bodies;
- Nurturing of new, genetically modified biological species;
- Development of important new drugs;
- Control and treatment of major contagious diseases such as AIDS and viral hepatitis;
- Large aircraft;
- High-resolution earth observing system;
- Manned space flights; and
- Lunar exploration projects.[10]

The central government has committed to releasing 99 policies elaborating goals of the Long-Term S&T Plan by the end of 2007 and has already released several of these documents. A sampling indicates the breadth of their coverage:[11]

---

[10]Ibid.

[11]*Guo Ban Han [2006] No. 30* (A Letter in Reply from the General Office of the State Council on Approving the Formulation of the Rules for Implementation of the Several Supporting Policies for Implementation of the Outline of the National Medium and Long-term Plan for Development of Science and Technology), *Gazette of the State Council,* Issue No. 17 Serial No. 1196, June 20, 2006. Available at <*http://www.gov.cn/gongbao/content/2006/content_310755.htm*> (provisional translation on file).

So far, 5 of the expected 99 documents to be issued in support of the Long-Term S&T Plan are known to be published. The full set of 99 is expected to be published by the end of 2007.

The titles and subject areas for the five published are listed below:

1. *Guidelines for Priority Areas of High-tech Industrialization and Development, June 2006,* most recent and is sixth on the State Council's list of those 99; underscores the importance of focusing on industries with independent innovation.

2. *Provisional Regulations on the Management of State Projects of High-tech Industry Development, February 28, 2006.* This document provides general guidance on the organizational structure and management operations of national high-tech projects.

3. *Notice of the General Office of the State Council on Distributing Guiding Opinions on Building State Engineering Laboratories, July 13, 2006.* This document provides that a core group of new national engineering laboratories will be established, in part to reduce dependence on foreign technology.

4. *Measures pertaining to the Selection of Postgraduate Students Sent abroad by the State for Public Duties 2006, undated.* This document states that outstanding postgraduate students sent abroad will be identified and honored.

5. *Several Opinions of the Ministry of Commerce, National Development and Reform Commission [et al.] on Encouraging Technology Introduction and Innovation and Promoting Reform of Foreign Trade, July 14, 2006.* This document identifies nine industries in which foreign technology acquisition will be particularly encouraged.

- Accelerating Creation of Independent, 'Well-known' Chinese Brands;
- Supporting Technology Innovation of Small and Medium-sized Enterprises;
- Issuance of Corporate Bonds for Qualified High-Tech Enterprises;
- Regulation on Management of Start-up Investment Funds and Debt Financing Ability of Start-ups;
- Suggestions on Establishing and Improving Regional Intellectual Property;
- Standardizing Foreign Acquisition of Key Chinese Enterprises in the Equipment Manufacturing Industry;
- Building Research-Orientated Universities;
- Promoting the Development of State Supported High and New Technology Industry Development Zones;
- Establishing Guidelines and Funding for Venture Capital Investment;
- Creating Tax Policies Supporting the Development of Start-ups; and
- Establishing "Green Channels" for High-level Talents Who Have Studied Abroad to Return to China.

The breadth and depth of this undertaking are massive by any measure and are designed to equal results achieved by further evolved market economies that had a head start of decades and, in some cases, more than a century. This requires China to acquire a financial, educational, and legal infrastructure quickly to support an economy whose growth is based on innovation.

WHY INNOVATE?

The motivation for this grand effort was described in National Development and Reform Commission Minister Ma Kai's recent remarks:[12]

> *. . . [W]e will promote development by relying on enhancing independent innovation capability, take it as a national strategy and shift economic growth from relying on the input of capital and material factors to relying on scientific and technological advancements and human resources. This is a solution put forward to address the bottlenecks of science and technology and human resources China encounters in economic and social development.* ***China's economic growth largely relies on material inputs and its competitive edge is to a great extent based on cheap labor, cheap water and land resources and expensive environmental pollution. Such a competitive edge will be weakened . . . with the rising price of raw materials and the enhancement of environmental protection. Therefore, we should enhance [our] independent innovation capability . . . and increase the contribution of science and technology advancement[s] to [our] economic growth. . . .***

[12] Ma Kai, Minister, National Development and Reform Commission, The 11th Five-Year Plan, March 19, 2006, available at *<http://english.gov.cn/2006-07/26/content_346731.htm>*.

In fact, a primary emphasis of the 11th Five-Year Plan period is to "give more strategic importance to developing education," creating a large number of "highly talented people [to build] . . . a strong nation with abundant human resources."

INNOVATION POLICIES

Intellectual Property

In the course of being developed is also the National Intellectual Property Rights strategy. Lu Wei, deputy director general of the Technical Economic Department, Development Research Center of the State Council, described the policy in the following terms:[13]

> *To adapt the Strategy to China's development situation, there are three fields that we shall pay attention to:*
>
> *First, [we shall] abide by international principles and meet the lowest protection standards of the WTO, and offer reasonable protection.*
>
> *Second, [we shall] not only encourage self innovation, but also encourage absorption, consumption, and innovation of introduced technologies and pay attention to IPR protection of technological transactions and technological licenses.*
>
> *Third, [we shall] strengthen the antimonopoly [activities] related to IPR and prohibit multinationals from shutting domestic enterprises out of the market using IPR.*

This statement is of interest at various levels. It is considered important by China's policymakers that China protect intellectual property, because most IP is foreign, and there will be no opportunity to attract and absorb foreign technologies without the protection of intellectual property. However, there is also a strong sense that excessive dependence on foreign technology is harmful to China. Therefore, adherence to "the *lowest protection* standards of the WTO [World Trade Organization" allows China to honor the letter of its WTO commitments while protecting and preserving China's domestic markets for domestically produced "innovation" goods.

The schizophrenia over the need for and reaction against the influx of foreign technology is also evident in discussion of potential for use of China's new antimonopoly law (discussed further below) to impede competition from foreign companies with strong patent portfolios for the benefit of domestic competitors.

---

[13]Wang Changyong, "IPR Sails Against Current Stream," *Caijing Magazine,* October 17, 2005. Available at <*http://caijing.hexun.com*>.

Standards Policy

A tool China intends to use to promote indigenous innovation (and restrain foreign competition) is standards setting. The recently promulgated Shanghai Municipal Government Intellectual Property Strategy demonstrates how standards setting can be used to foster innovation. The Shanghai Strategy states:[14]

> *[We shall] actively promote the formulation and implementation of technical standards with self-owned intellectual property rights and translate that technological advantage into a marketplace advantage to maximize the benefits of intellectual property rights.*

This sentiment is carried one step further in the National Medium and Long Term Program on Scientific and Technological Development.[15]

> *[We shall] actively take part in the formulation of international standards, and drive the transferring of domestic technological standards to international standards. . . .*

The recent and highly contested case of the WAPI wireless standard is an example of this policy in practice. With WAPI, China sought to impose an indigenous, not internationally recognized standard for wireless networking within China. Although China was entirely within its rights to impose this standard for domestic consumption, the plan would have required each foreign supplier of wireless computers to share its proprietary technology with a Chinese partner company. In turn, the Chinese partner would supply an essential encryption algorithm, without which the product would have been denied access to the Chinese market. After receiving strong representations from Colin Powell and other U.S. cabinet secretaries, the plan was put on hold indefinitely. That was not the end of WAPI, however, as China has attempted to have its WAPI domestic standard recognized and accepted at WIPO (the World Intellectual Property Organization). As yet, little progress as been made by China toward attaining that objective.[16]

---

[14]Shanghai Municipal Government, *Notice of the Shanghai Municipal Government Regarding Distributing the Outline of Shanghai's Intellectual Property Strategy (2004-2010),* September 14, 2004. Provisional translation on file; original available at <*http://www.sh.gov.cn/. . .*>. Shandong province has issued a similar policy statement. See *People's Government of Shandong Province, Outline of Shandong's Intellectual Property Strategy,* July 14, 2005. Provisional translation on file; original available at <*http:/www.sdpatent.gov.cn/news/gangyao.htm*>.

[15]State Council of the People's Republic of China, *Outline of the National Medium-and Long-Term Program on Scientific and Technological Development (2006-2020),* February 9, 2006. Provisional Translation from Chinese on file; original available at <*http://www.cas.cn/html/Dir/2006/02/09/13/70/88.htm*>.

[16]Yang lei, ed., "Chinese WAPI Delegation Quits Prague Meeting," Xinhua, June 8, 2006. Available at <*http://news.xinhuanet.com*>.

Government Procurement

Another policy instrument used by China to seek to foster innovation is government procurement. Government procurement has long been a tool to provide preference and protection to developing industries. The WTO's Government Procurement Agreement is designed to ensure that member governments utilize open, transparent, competitive, unbiased, merit-based, and technologically neutral procurement procedures. In 2001 China committed to initiate negotiations for membership in the GPA "as soon as possible,"[17] and at the April 2006 U.S.–China Joint Commission on Commerce and Trade (JCCT) meeting stated: "Negotiations on China's entry to the GPA will be launched no later than the end of December 2007."[18] However, recent Chinese policy documents indicate that China intends for its state institutions to go against the basic tenets of the GPA:

> *We shall . . . introduce an open, fair and square competition mechanism to embody preferential policies for government procurement. . . .*
>
> *Finance departments at provincial level shall work with the science and technology departments at the same level to establish implementation plans for developing indigenous innovation government procurement policies for their provinces.*[19]

Due to the prominent role China's centralized governmental structure plays in China's economy, the use of a discriminatory government purchasing policy at the central, provincial, and local levels can provide a significant amount of protection to foster indigenous innovation and may have a very powerful negative effect on trade. The State Council has decreed:

> *The government shall set a priority procurement policy on important high-tech equipment and products developed by domestic enterprises with independent intellectual property. [We shall] provide policy support to enterprises purchas-*

---

[17]Working Party on the Accession of China, *Report of the Working Party on the Accession of China* WT/MN(01)/3, November 10, 2001.

[18]Transcript, Press Conference with U.S. Commerce Secretary Carlos Gutierrez, U.S. Secretary of Agriculture Mike Johanns, U.S. Trade Representative Bob Portman, Chinese Vice Premier Wu Yi at the Annual Meeting of the U.S.–China Joint Commission on Commerce and Trade, The Department of Commerce, Washington, D.C., April 11, 2006. Available at *<http://www.ustr.gov/>*. See also Inside US-China Trade, Industry Worried China Backing out of Commitment to Join GPA, September 27, 2006.

[19]Chinese Ministry of Finance, *Opinions of the Ministry of Finance on Implementing Government Procurement Policies That Encourage Indigenous Innovation,* Cai Ku [2006] No. 47, June 13, 2006. Provisional Translation from Chinese on file; original available at *<http://www.zjzfcg.gov.cn/jcms_public/jcms_files/jcms1/web1/site/art/2006/06/28/art_103_976.html>*.

ing domestic high-tech equipment, and support the formulation of technological standards through government procurement.[20]

One of the government procurement policies that China's trading partners have found to be most objectionable is the policy on procurement of software.[21]

Fostering Inward Foreign Direct Investment (FDI)

Currently, China is heavily dependent upon foreign direct investment (FDI) as the primary source for innovative capacity building within China. According to China's Ministry of Commerce—

[Foreign direct investment]...is an important element of China's fundamental principle of opening up to the outside world, and also an important component of Deng Xiaoping Theory, and is one of the great practices of building up [a] socialist economy with Chinese characteristics.[22]

As a result many of the policies supporting innovation also seek to attract "high end" FDI. To encourage foreign investment in R&D, tax incentives are granted. For example:

- Income earned from the transfer of technology developed exclusively through foreign invested R&D is exempt from sales tax.

- Foreign R&D investors with development expenses at least 10 percent greater than year previous expenses are entitled to a 50 percent discount in total technological development expenses in the current year corporate income tax.[23]

---

[20]State Council of the People's Republic of China, *Outline of the National Medium- and Long-Term Program on Scientific and Technological Development (2006-2020),* February 9, 2006. Provisional translation from Chinese on file; original available at <*http://www.cas.cn/html/Dir/2006/02/09/13/70/88.htm*>, p. 21.

[21]China Trade Extra, *China Agrees to Delay Software Procurement Rule While Talking with U.S.,* July 11, 2005, available at <*http://www.insidetrade.com*>, and The U.S.–China Joint Commission on Commerce and Trade (JCCT), *Outcomes on Major U.S. Trade Concerns,* The Office of the United States Trade Representative, July 7, 2005, available at <*http://www.ustr.gov*>.

[22]Ministry of Commerce of the People's Republic of China, *Overview of FDI in China.* Available at <*http://enlgish.mofcom.gov.cn/*>.

[23]Guogiang Long, "China's Policies on FDI: Review and Evaluation," in *Does Foreign Direct Investment Promote Development?,* Theodore H. Moran, Edward M. Graham, and Magnus Blomström, eds., Institute for International Economics, May 2005.

Import Policy

While foreign investment is attracted and managed, trade policy is not neglected. Although it is not yet clear how this will be accomplished—given China's WTO commitments not to interfere with imports—trade policy is to be utilized to foster innovation policy. The Ministry of Commerce recently issued the following policy pronouncement:

> *[We shall] . . . optimize the structure and quality of technology import[s] . . . to increase the investment in assimilation and absorption of . . . imported technologies to gradually establish a market-oriented system of technology imports and innovation. . . .*[24]

There are other references to import policy that give an indication of the direction of current thinking. One of the first sectoral Opinions of the State Council, this one relating to the "equipment manufacturing industry," states that "imports of key equipment using foreign capital will be subject to "strict examination and study."[25] Given WTO commitments, there is little that can be done through direct controls to slow the inflow of undesired imports (outside of the government procurement measures mentioned above). However, whenever the state has a benefit to confer or deny, there is the possibility to influence the kind and quantity of imports brought in by a particular investor. This may be what is intended, as the *Opinion* goes on to state that China's government will support "reorganization of different industries, regions and ownership on the basis that the state will retain the control and the dominant authority."[26] It also requires large-scale and technologically important equipment producers to "seek the opinion of the State Council's relevant departments if they want to sell stakes to foreign investors."[27]

The stated goal of this State Council Opinion is to establish by 2010 competitive Chinese equipment manufacturing companies with their own intellectual property to meet China's needs in energy, transportation, raw material, and defense sectors.

---

[24]Press Release, Ministry of Commerce of the People's Republic of China, *8 Ministries Including Ministry of Commerce putting Forward Policies on Promotion of Technology Import and Innovation,* August 3, 2006. Available at *<http://english.mofcom.gov.cn/aarticle/newsrelease/significantnews/200608/20060802789109.html>.*

[25]China State Council, *Several Opinions of the State Council on the Acceleration of Revitalization of the Equipment Manufacturing Industry,* February 13, 2006. Provisional translation on file; original at *<http://www.ndrc.gov.cn/gyfz/zcfg/t20060626_74596.htm>.*

[26]*Supra* Note 21 and Guo Xinxyu and Zhang Qinyuan, "Three Barriers in the Equipment Manufacturing Industy," *Economic Observer,* May 15, 2006. Available at *<http://www.eeo.com.cn/>.*

[27]Ibid.

Competition Policy

As noted briefly earlier, another policy tool which may be employed by China to foster innovation is its new (and as yet unimplemented) antimonopoly law (AML). Competition laws have served various purposes in various jurisdictions where they exist. In the United States, Canada, the United Kingdom and Germany, the stated goal has been primarily the protection of the consumer. In other places, economic development, industrial policy, or social objectives may be primary policy drivers.[28] In China, there is clearly a concern by top-level policymakers that an imbalance between China's indigenous companies' portfolio of intellectual property rights and those of its trading partners' companies can be viewed as being highly problematic. Wang Xiaoye, a professor of Anti-Monopoly Legislation at the Chinese Academy of Social Sciences, noted that China's AML will create an environment of fair competition within China, while also referring to the problem that multinational companies possess capital and technological advantages allowing them to dominate the market "more quickly."[29] With this in mind, she concludes that "[t]he adoption of an anti-monopoly law will serve as an important tool for China to check the influence of multinationals."[30]

Reinforcing this view, in 2004 the Fair Trade Bureau of the State Administration for Industry and Commerce (SAIC) released a report claiming that certain multinational companies utilize technological advantages and IPR to dominate sectors of the Chinese market.[31] The report specifically names Kodak, Tetra Pak, and Microsoft as both examples and potential targets of any forthcoming Chinese Anti-Monopoly Law.[32] Additional press reports indicate that other advanced technology and innovative companies such as Intel could also be a target of such legislation.[33]

---

[28]When Japan liberalized inward FDI in the 1960s, policymakers regarded the Antimonopoly Law as a foil to prevent the feared "monopolization" of the domestic market by "foreign capital." A Ministry of Finance official serving on the Foreign Investment Deliberation Council, which was developing Japan's FDI policy, commented "When Foreign capital comes out strong, not only in terms of technological or capital strength but starts stirring up the market using violent force, I believe there is a strong underlying desire by all to use the Antimonopoly Law to police illegal transactions." Fujio Yoshida in "Preparation of Legal System for Capital Liberalization (Part 3)," Panel Discussion in *Zaikei Shoho,* July 17, 1967.

[29]Dai Yan, "Anti-Monopoly Legislation on the Way," *China Daily,* June 18, 2004.

[30]Ibid.

[31]Office of Anti-monopoly, Fair Trade Bureau, State Administration of Industry and Commerce, "Anticompetitive Practices of Multinational Companies in China and Countermeasures," *Journal of Administration of Industry and Commerce,* May 2004. Provisional translation on file. See also Nathan Bush, "Chinese Competition Policy, It takes more than a law," *China Business Review,* May-June 2005; Dai Yan, "Anti-Monopoly Legislation on the Way," *China Daily,* June 18, 2004; and "Nation May Introduce Antimonopoly Law," *Shenzhen Daily,* December 30, 2005.

[32]*Supra* Note 3 *Anticompetitive Practices.*

[33]"Nation May Introduce Antimonopoly Law," *Shenzhen Daily,* December 30, 2005.

The June 22, 2006, circulated draft of the Anti-Monopoly Law illustrates how this law may be potentially used to support domestic development and/or restrict competition. For example, Article 10: Exemptions of Monopoly Agreements establishes exemptions to the draft prohibitions on horizontal and vertical monopolies if the "monopoly agreements" meet objectives such as: improving operational efficiency and increasing competitiveness of small and medium sized enterprises; "ensure the legitimate interests in foreign trade and economic cooperation"; reduce costs; raise product quality and efficiency; and harmonize specifications and standards for specific products while still benefiting consumers and "not substantially eliminat[ing]" competition in the market.[34]

A key concern will be how "monopolistic conduct" which includes "abuse of dominant market position" is defined.[35] Article 14: *Direct Determination of Dominant Market Position* establishes that entities within a "relevant" market are considered to hold a dominant market position if they have any of the following conditions:

> *(i) The market share of one undertaking in relevant market[s] accounts for more than 1/2;*
> *(ii) The joint market share of two undertakings as a whole in relevant market[s] accounts for more than 2/3;*
> *(iii) The joint market share of three undertakings as a whole in relevant market[s] accounts for more than 3/4.*[36]

Of course, much will depend on the defining the relevant market. In China, a foreign company with a strong patent portfolio might easily command a large portion of a product market. As the Medium and Long Term S&T Plan states:

> *[We shall] prevent the abuse of intellectual property that unfairly restricts the market mechanism for fair competition and may prevent scientific-technological innovation and the expansion and application of scientific-technological achievements.*[37]

Foreign companies have already been put on notice that the new antimonopoly law (AML) might find tempting targets of Western companies that have the strongest IP positions—such as Microsoft, Tetra Pak, Cisco, and Kodak.[38]

---

[34]*Article 10: Exemptions of Monopoly Agreements,* Anti-Monopoly Law of the People's Republic of China, Revised July 27, 2005, as circulated on July 27, 2005, unofficial translation on file.

[35]See *Supra* Note 6 Article 3.

[36]*Supra* Note 5 Article 14.

[37]State Council of the People's Republic of China, *Outline of the National Medium-and Long-Term Program on Scientific and Technological Development (2006-2020).*

[38]"Anticompetitive Practices of Multinational Companies and Countermeasures," Administration of Industry and Commerce, Office of Anti-monopoly, Fair Trade Bureau, State Administration of Industry and Commerce, May 2004, pp 42-43.

Direct Funding

Direct government financial support is also an important part of Chinese government innovation policy as it is for all countries seeking to promote innovation. Thus, for the most part, China is not out of step with its competitors in seeking to promote indigenous technological development through government funding. One example from the current Chinese experience consists of the measures to promote the development of a domestic equipment manufacturing industry, measures which include:[39]

- Preferential taxation;[40]
- Incentives for purchase of Chinese-made machinery;[41]
- VAT rebates on imported parts and materials;
- Allocation of special funds for technologically advanced products;[42] and
- Relief of enterprises' "social responsibilities."

While the phrase "relief of enterprises of social responsibilities" has not been defined, American automobile producers (and former executives of former steel manufacturing companies) can testify that if this means "legacy costs," that is, health and pension costs of retired workers, among other "social responsibilities," relief of these costs can mean the difference between profitability, stunning losses, and in the extreme, but all too well known cases, bankruptcy.[43]

By far the highest profile recent instance of government funding is the announced agreement by two municipalities to attract semiconductor fabrication facilities through 100 percent funding of capital costs of a private company. The beneficiary is SMIC (the Shanghai Manufacturing International Corporation), a major semiconductor foundry. SMIC has announced that it will receive the benefit of construction of two new chip fabrication facilities, from Chengdu and Wuhan

[39]State Council, *Several Opinions on the Revitalization of the Equipment Manufacturing Industry,* published on June 28, 2006.

[40]The Ministry of Commerce (MOFCOM), NDRC, and relevant state organs are to implement specific preferential tax policies, provide exemptions from tariffs, and provide value-added tax rebates with respect to the importation of key parts and raw materials used to develop key equipment and projects.

[41]Companies are encouraged to purchase their "initial" equipment from domestic manufacturers. Purchasers, equipment makers, and insurance companies are to work together to share risks and profits arising out of such purchases. Insurance companies will be guided to provide insurance coverage to the companies making domestic purchases.

[42]Increasing financial support to key technological equipment manufacturers. The government will allocate special funds annually for the creation of key programs and technologically advanced projects.

[43]Relieve enterprises' societal responsibilities. National and local governments are to provide funds and support to relieve enterprises of their "societal responsibilities." The state-owned Assets Supervision and Administration Commission of the State Council are to facilitate this process and reduce social burdens on enterprises. *China State Council Issues Opinions on Equipment Manufacturing Industry,* Dewey Ballantine LLP, available upon request.

(and Hubei) local government agencies. This amounts to a grant equivalent to billions of U.S. dollars. SMIC is to receive an operating contract and the plants are dedicated for the sole use of SMIC. According to press reports, SMIC will also receive a "management fee" and will have the option to buy the plants in the future. Current profits from operations would be retained by SMIC. This goes beyond most examples of innovation funding as far as I know and is one of the most extreme cases.[44]

High-tech Incubation Parks

Information on company-specific benefits is not usually publicly available to this extent, especially in the case of state-owned enterprises. A more common, historic, and well-documented form of support for private enterprise is the use of industrial incubation parks, which in China are built on a scale not seen anywhere else on the planet. The parks are the showcases and a chief element of pride of Chinese government innovation planners. As Ministry of Science and Technology official Xu Lupin said, "China considers science parks to be central to its efforts to build capacity for innovation. . . ."[45] Thirty more parks are planned to be established by 2010. By 2004, parks had accounted for 38,565 participating companies, US$ 226.4 billion of production, and US$19.7 billion investment in infrastructure.[46]

---

[44]The business model includes an option to buy fabs in 3-5 years "at original price." "[w]e have 100% of the authority to operate and to manage the fab, and to fully use that capacity." ["Semiconductor Manufacturing International Q1 2006 Earnings Conference Call," April 27, 2006.] "[t]he profit from this operation belongs to us, and we will pay whatever we need to the fab owners. So this is the method. Almost like we rent the fab, almost like that. It's basically outsourcing a model."["Semiconductor Manufacturing International Q1 2006 Earnings Conference Call," April 27, 2006.]

Wuhan Xinxin Semiconductor Manufacturing Corp. SMIC announced at the end of June that it is participating in a multi-billion dollar fab investment being made by the Hubei provincial and Wuhan city governments. "Chinese Government Builds 300mm Fab," Semiconductor Fabtech, June 28, 2006; "Chinese Province Pays to Get 300-mm Wafer Fab," EE Times, June 28, 2006; "Chip Giant Quickens Steps of Expansion," TMCNET, June 28, 2006.

For the time being, the entity is fully government-owned and financed by an investment company associated with the Hubei Provincial Government, Wuhan City Government, and the Wuhan East Lake New Technology Development Zone. SMIC describes its participation as a management contract under which SMIC "will not invest any money to construct or equip the wafer manufacturing facilities but will manage the operations, including the wafer loadings, of the facilities."[*Semiconductor Manufacturing International Corp. Form 20-F for Period Ending December 31, 2006,* p. 23.] SMIC reportedly signed an MOU in conjunction with this contract in March 2006. ["SMIC Gets $3B Nod from Chain's Wuhan Government," *Electronic News,* May 22, 2006.]

The SMIC example may well not be out of the ordinary in China where State-owned enterprises account for a substantial portion of the economy.

[45]Wu Ching, China to Build 30 new science and technology parks, SciDev.net, April 19, 2006, available at *<http://Scidev.net>*.

[46]Dr. Yong Shang, Vice Minister Science and Technology China, Speech, "Innovation: New National Strategy of China," as presented at Levin Institute Conference, "Industrial Innovation in China," July 2006.

A good example is the Tianjin Binhai New Area, which claims investments from 69 companies of the top Global 500 and 42 R&D centers. On May 26, 2006, the State Council declared the project of "national importance." The purpose of the Binhai Hi-Tech Industry Park is to import S&T personnel, R&D institutions and "knowledgeable enterprises," and provide them with support services for patent applications, investments, finance, and corporate regulation. The stated goal is to have the Binhai New Area become the "medicine port" for East Asia.[47]

As is often the case in China, different areas are in sharp competition with each other to incubate new industries and foster innovation. This has been the case with Shanghai and Beijing for some time, with major bids being made by Wuhan and Chengdu and numerous others.

Shanghai Zhangjiang Hi-Tech Park aspires to become both China's Silicon and Pharmaceutical Valleys. China's "Pharmaceutical Valley," occupying 16 square kilometers is "to form a perfect Hi-tech innovation chain," having attracted $10.6 billion in foreign capital, including facilities of 42 foreign companies including Roche, GlaxoSmithKline, and Medtronic. It has established 31 R&D institutes, and has provided a hospital for clinical trials.[48]

Shanghai Zhangjiang Hi-Tech Park, in its desire to become "China's Silicon Valley," states that it has attracted 70 fabless[49] companies, 3 foundries, 2 photomask producers, 12 packaging and test companies, 34 equipment vendors, and numerous systems application companies.

A common thread throughout the various initiatives is the Chinese planners' emphasis on absorption of foreign investment. It is, as the Ministry of Commerce stated in 2004—

> *"an important part of China's fundamental principle of opening up to the outside world, and an important component of Deng Xiaoping Theory. [It] is one of the great practices of building up socialist economy with Chinese characteristics.*[50]

The inward flow of foreign direct investment, not only to attract manufacturing for its many benefits, but for technology transfer, is vital. Particular emphasis is given to attracting foreign R&D facilities. Fifteen Korean companies have R&D

[47]Professor Li Jianjun, Director General, Tianjin Science and Technology Commission, "The Development and Opening of Tianjin Binhai: New Area & China's Biotechnical Innovations" as presented at Levin Institute Conference, "Industrial Innovation in China," July 2006.

[48]*Shanghai Zhangjian Hi-Tec Park: China's Silicon Valley and Pharmaceutical Valley* at *<http://www.localglobal.de/gbf2004/vortraege/shanghai_zhangjiang.pdf#search=%22Shanghai%20Zhangjiang%20Hi-Tech%20Park%20%22>* and *<http://www.chinatoday.com.cn/English/>*.

[49]Fabless semiconductor producers are design houses that have their products "fabbed," that is produced, by a company having a fabrication facility.

[50]Ministry of Commerce, June 14, 2004, available at *<http://www.fdi.gov.cn/Iteconomy/>*.

centers in China, 14 of these established since 2000. Samsung and LG Electronics have three each. These concentrate on development of technology and product models for the Chinese market.[51]

FACTORS DRIVING AND INHIBITING INNOVATION

Major Positive Factors

Innovation efforts in China are a vast work in progress, conjuring up images of past grand public works efforts—more like perhaps the building of Grand Canal system[52] than the Great Wall, but there are more than a few trace elements of the latter, protective element, that appear in the statements of Chinese planners and regulators.

Growth

There are strong innovation drivers in China at present. First among these is an impressive rate of GDP growth. While this can be something of a double-edged sword, in that resources can be scarce in a rapidly growing economy, clearly the positives far outweigh the negatives—for those who seek rewards for innovation, more are to be found in a buoyant economy.

Human Resources

A second innovation driver is the huge talent pool. Enormous resources are being poured into graduating engineers and scientists, and given the immense population base, this is an effort that can in sheer numbers equal the output of many of China's foreign competitors, taken together.

Market Size

The large domestic market is both an incentive to indigenous Chinese production and also serves as a magnet for foreign direct investment. It is a market that a global company cannot afford to ignore. And it is from these foreign companies that potentially much can be learned, leapfrogging the painful earlier steps in innovation that were required of the technology donor companies. It helps to attract higher end foreign investment that the Chinese market is increasingly a

---

[51]Professor Yong-June Kim, School of Business and Dir. China Research Institute, SKK University, Seoul, Korea, "A Korean Perspective on China's Innovation System," as presented at Levin Institute Conference, "Industrial Innovation in China," July 2006.

[52]Conceived of by the Duke of Wu in the eighth century B.C., but really culminating in the construction during the SUI dynasty some 1,100 years later.

very sophisticated one. Cell phone and computer penetration in China are very high.

Increasing Protection of Intellectual Property

Formal intellectual property protection, poorly developed, to non-existent, in much of China until relatively recently is making strides, particularly in Beijing and Shanghai. There have been a number of cases which have resulted in satisfactory outcomes for foreign and domestic IPR holders in China. While there is much progress to be made, it is clear that intellectual property protection is improving. This movement is being bolstered by the incentives for indigenous patenting which create domestic stakeholders in a functioning IPR system.

Setting Priorities

On top of these advantages is a government (or more accurately, a series of governments, at the central, provincial, and municipal levels) pledged to full economic mobilization to support innovation—and at the provincial and local levels in a rivalry to achieve often grand objectives. The setting of priorities may be particularly effective where their objectives are specific:

> *[We will] significantly increase* ***the self-sufficiency ratio to over 70 percent for integrated circuits used for information and national defense security, and to over 30 percent for integrated circuits used in communications and digital household appliances....*** *We should basically achieve self-sufficiency in key products supply....*[53]

Some Indications of Progress

There are a variety of concrete measures of success. China's high-tech exports have been growing at an annual rate of more than 40 percent over the last 5 years. China is now third in the ranking of countries in R&D expenditures. Another crude measure of success is the number of domestic patents filed with China's State Intellectual Property Office (SIPO): 99,278 patents in 2001, 171,619 in 2005, a 73 percent increase.[54] Whether this represents true innovation will only be a matter to be assessed with hindsight—as the high-tech exports are

[53] "Outline of the 11th Five-Year Plan and Medium-and-Long-Term Plan for 2020 for Science and Technology Development in the Information Industry," Ministry of Information Industry, Xin Bu Ke [2006] No. 309, August 29, 2006. Note that the text of the Outline itself is undated, but the MII Web site posting the Outline is dated August 29, 2006.

[54] China's State Intellectual Property Office listings for 2001-2005, available at <*http://www.sipo.gov.cn/sipo_English/statistics/200607/t20060725_104687.htm*>. Listed patents included Innovation, Utility Models, and Design.

probably very heavily the products of foreign multinational corporations and their Chinese joint ventures.

Some Negatives

Other countries' innovation planners must accord China a measure of respect for its accomplishments to date, given the very low level from which it started so very recently. But imbedded in some of the very causes of success are also issues that can and do serve to inhibit innovation. Among these are the major overhang of the vestiges of a command economy and the newness and in some instances lack of the legal and market infrastructure characteristic of a nonmarket form of economic organization.

Government as a Mixed Blessing

State planning is a double-edged sword. Those who make plans are not necessarily those closest to the cutting edge of innovation and may not fully understand its needs. Nonmarket factors may skew economic activity. When the cheering died down a bit, observers of the Japanese miracle began to note that Japan suffered deeply from problems of crony capitalism that saddled its banking system with nonperforming loans and contributed to depressing its economic growth for over a decade. In China, some two-thirds of the economy is accounted for by state-invested companies and state-owned companies. No one has accused these companies as a group of being on the leading edge of innovation. The influence of the state can be too pervasive. Complicating the positive story of the dominance of market forces today are stories of the recent resurgence of the Communist Party's involvement in business. What impact this will have it is too early to tell, but it is likely to reinforce a relationship-based pattern of transactions that may often run counter to the dictates of the market.[55]

Workforce Issues

In the rest of the world, particularly in the United States, Europe, and Japan, but also in India, and others of China's competitors, the number of China's engineering graduates graduated annually is causing concern. Newspaper articles have

---

[55]*. . . The Communist party is insinuating itself more deeply into business by setting up committees in many of the country's 80m-odd private companies. . . . Its economy may have been opened up, rigid central planning phased out, competition unchained and shares in state companies sold off. But critical levers of control remain firmly in official hands. Do you need a bank loan, foreign exchange, a site for a new plant or access to scarce raw materials? Then, in China, you need friends in the right places*. Guy de Jonquieres, "China's Curious Marriage of Convenience," *Financial Times*, July 19, 2006.

reported that the 70,000 U.S. science and engineering graduates for 2004 pales in comparison to China's 600,000.[56]

But, those numbers of graduates do not tell the whole story. A number of studies have found these comparisons to be misleading. A study by Duke University found the United States actually produced 137,437 while China produced 351,537 graduates in the Engineering, Computer Science, and Information Technology fields.[57] The study largely attributes the disparity in these numbers to issues of quality. China's state-centered and rote learning approach to education is heavy on theoretical and Marxist learning, producing "ivory tower" engineers with little problem solving and teamwork skills.[58] Engineering curricula are often made more crowded with ideological courses that detract from the quality of the graduates entering the workforce. China's educational system, Marxism included, is said to be inadequate, emphasizing depth over breadth, quantity of graduates over quality, leading to "transactional vs. dynamic engineers," and since it does not nurture creativity, producing an insufficient number of "innovative" engineers.[59] The Ministry of Science and Technology, and other planners are not unaware of these defects and are issuing new state guidelines and opinions to ". . . further strengthen the cultivation of talents in short supply. . . ."[60]

Inadequate Protection of Intellectual Property Rights (IPRs)

A key roadblock to accelerating innovation is the inadequate protection of intellectual property. This exists in part because there is relatively little history or culture of protecting intellectual property rights and there is only a recent history

---

[56]*Framing the Engineering Outsourcing Debate: Placing the United States on a Level Playing Field with China and India,* Duke University Master of Engineering Management Program. Available at <*http://memp.pratt.duke.edu/downloads/duke_outsourcing_2005.pdf#search=%22Duke%20study%20engineer%22*>.

[57]Ibid.

[58]Jo Johnson and Richard McGregor, "Are India and China up to the Job?," *Financial Times,* July 12, 2004. See also Guy de Jonquieres, "To Innovate, China Needs More than Standards," *Financial Times,* July 12, 2004, and Guy de Jonquieres, "China and India Cannot Fill the World's Skills Gap," *Financial Times,* July 12, 2004.

[59]*Framing the Engineering Outsourcing Debate: Placing the United States on a Level Playing Field with China and India,* Duke University Master of Engineering Managmenet Program. Available at <*http://memp.pratt.duke.edu/downloads/duke_outsourcing_2005.pdf#search=%22Duke%20study%20engineer%22*>. Cong Cao and Denis Simon, "China's Evolving S&E Talent Pool and Its Roles in Industrial Innovation," presentation at Levin Institute Conference, "Industrial Innovation in China," July 25, 2006. McKinsey Global Institute, *The Emerging Global Labor Market,* June 2005, available at <*http://www.mckinsey.com/*>.

[60]*Guo Ban Han [2006] No. 30* (A Letter in Reply from the General Office of the State Council on Approving the Formulation of the Rules for Implementation of the Several Supporting Policies for Implementation of the Outline of the National Medium and Long-term Plan for Development of Science and Technology), *Gazette of the State Council,* Issue No. 17 Serial No. 1196, June 20, 2006, available at <*http://www.gov.cn/gongbao/content/2006/content_310755.htm*> (provisional translation on file).

of private property.[61] The result is that China's share of world patents is very low, and Chinese officials state[62] that the quality of patents is also low with some 99 percent of Chinese firms owning no patents. The situation may be changing for the better. The State Intellectual Property Office (SIPO) has provided figures on shares of Chinese patents that do show at least a quantitative improvement:

- 2002: foreign 73 percent, Chinese 27 percent.
- 2005: foreign 54 percent, Chinese 46 percent.

IPR enforcement is going to be a problem for some time to come. With the best enforcement will in the world, there is and will remain a serious shortage of IP specialists, both legal and management. There is also a basic domestic public policy question involved in accelerating IPR enforcement. Counterfeit goods are far less expensive than branded items. With a low per capita GDP and a great divide between those with disposable income and those in poverty, the balance of the goals of enhanced IPR enforcement vs. some alleviation of poverty can weigh in favor of the latter.

Forced Technology Transfer

Lastly, there is an element of IPR that is prevalent but on which statistics will never be available, and that is forced technology transfer. In any governmental regime, but particularly in the Chinese case, where there is a strong desire on the part of foreign firms to enter the market to sell and/or to produce, the price paid for entry can and very often is an agreement on the kind and amount of technology that will be brought into China. While not quantifiable, issues of this kind surface from time to time that indicate that the problem is extensive and important. One example is the nonpayment of royalties to Japanese manufacturers for DVD players. This is a very large market segment, and the royalties owed can be assumed to have been very large. They are involuntarily waived. There is a view in the industry that official administrative guidance was given that royalties should not be paid (and certainly there are many general pronouncements that the payment of royalties by Chinese companies should be avoided wherever possible). The practice is tolerated

---

[61]China's Westernization Movement (Yangwu Yundong), which arose in the 1860s, encouraged study of Western systems of science and technology, and in 1896, China's leading newspaper, Shen Bao, proposed a Western-style patent system. In 1898 China enacted a patent law and a Chinese inventor was awarded a patent for a novel form of spinning machine. However, decades of war and revolution followed, and under Mao, the government took the position that all intellectual property belonged to the state. Liwei Wang, "The Chinese Traditions Inimical to the Patent Law," *Northwestern Journal of International Law and Commerce,* Fall 1993.

[62]Dr. Hu Zhijian, Deputy Director General, Deptartment of Policy, Regulation and Reform, Ministry of Science and Technology, "IPR Policies In China: Challenges and Directions," presentation at Levin Institute Conference, "Industrial Innovation in China," July 2006.

because being in the Chinese market is profitable or is believed to be profitable in the future. Given what is at stake, the right to royalties is not surrendered by foreign companies. It is just not vigorously pressed.

Another IPR-related problem exists because of the very heavy dependence on inward foreign direct investment (FDI). Because of deficiencies in IPR protection, foreign investors withhold core technologies as well as the latest cutting edge technologies, thus limiting technology transfer to the more routine. Chinese planners are well aware of this problem and this spurs even greater efforts on their part to their fostering the creation of indigenous invention.[63]

The Regulatory Environment and Techno-nationalism

Lastly, a major inhibiting factor to innovation is a combination of negative (as seen from abroad) government promulgated measures. The Chinese government is far from monolithic, and this is not just a reference to the division of power with provincial and municipal governments. Within the Chinese government there are those who believe that the less intervention there is the better, and that the attractiveness of the market, left to some extent unfettered, is the best course for development. There are then those who believe in and practice a brand of techno-nationalism. The measures taken and threatened have been enumerated above, and include the whole panoply of investment controls, conditional financial support, possible administrative guidance to avoid payment of foreign royalties, threatened use of the antimonopoly law, regulatory approvals of all sorts, and the like.

As this is a paper focused on government measures, other market factors, such as lack of physical infrastructure, availability of feedstocks, and constraints caused by limited opportunities for synergies with local partner companies, will be reserved for other studies.

CONCLUSION

It would be clearly imprudent to draw sweeping conclusions about the rate of innovation in China at present. Extrapolation of trends of current efforts do not have very much predictive quality. In the 1980s it would have been and was a mistake to just extend the curve of then current Japanese growth. Similarly, Boeing presumably misjudged the prospective competition from Airbus. Most of China's innovation lies mainly in the future. What we do know are several factors upon which success could depend.

[63]Leonard Lynn, Case Western Reserve University, Hal Salzman, Urban Institute, "Collaborative Advantage and China's Evolving Position in the Global Technology System," as presented at Levin Institute Conference, "Industrial Innovation in China," July 2006.

The Gold Rush Attraction of the Chinese Market

We know that foreign firms still see China as a vast and vital market, as well as a production platform for export. While companies may be guarded in transferring technology, the greatest form of technology transfer is human, and there are a very large number of Chinese engineers working for foreign firms in China and those abroad who will repatriate. Stock options and other forms of economic and intangible rewards of participating in a new frontier as well as a better quality of life in some ways (a lower cost of living for one) act as magnets. U.S. immigration policy actually now pushes newly awarded PhDs from American universities to return home. Conditions are improving in all respects, and absent some upheaval, the polarity of the magnetism with respect to inward (into China) talent flows is unlikely to be reversed, just as it has been with respect to the movement of overseas ethnic Chinese back to Chinese Taipei in recent decades. In addition, the deficiencies in domestic Chinese training are known, and there is no reason to believe that they cannot be significantly rectified.

Remaining Uncertainties

There are, of course, many unknowns. A series of "ifs" exist—they always do. The move toward indigenous innovation will continue if political stability is maintained, if the instinct to techno-nationalism is checked (not manipulating trade, the antimonopoly law or product standards for mercantilist reasons, for example), if the inward flood of FDI remains largely unabated because the market continues to grow and China remains a low cost-production platform and the protection of IPR continually improves, and if the rate of economic growth continues to be strong, not cut off by either political or economic factors (such as weaknesses in the banking system, or external causes—reactions to trade imbalances, for example). In short, if the interventionists among the planners exercise enough self-restraint and support rather than acting to resist market forces and there are a few good breaks—such as continued global trade expansion—there are very good prospects for innovation taking hold in China in a very major way.

EPILOGUE

China's national innovation policy is a work in progress. Chinese officials have allowed much to take place that is healthy. Foreign firms' participation in the economy is essential, as is external collaboration by indigenous firms. To date the results have been mixed, but mainly positive: state intervention has its costs, but has been offset by welcoming a major wave of foreign investment.

The key question is: Can China have as much state-run intervention as it does at present and still create a market economy that maximizes innovation?

What is sure is that whatever is true today will change rapidly and will be different in a year, and perhaps unrecognizable in 5 years. Those of us who traveled to China in the 1980s and saw the empty fields of Pudong could not have foreseen that area of industrial and scientific ferment that exists today. The short survey above is partial, and for that reason cannot be predictive. There are areas not addressed at all, such as the role of universities, for the simple reason that their role was not stressed in most of the literature and discussions that went into the preparation of this piece. But the role of universities was important in the creation of Silicon Valley and Route 128, and may play a significant role in China's high-technology development. That subject alone deserves further study.

The emphasis in this review is on the role of the state. Part of that is due to the fact that the state is more dominant or at least more visible in China compared with other locations (even given the role of the U.S. Departments of Defense and Energy, and the National Institutes of Health in the United States, and various German, Japanese and French institutions and institutes). At least in its early stages, the role of the state in China is reminiscent of the historical role of the state in Japan through the Meiji Restoration. The relevance of Japan's experience is however limited by the very different path that China has taken, welcoming foreign investors and goods for its vast home market.

What is present in China is the will to succeed, and an excitement about China's growth potential and scientific and engineering possibilities that is somewhat reminiscent of the faith that the American settlers had in moving west in the mid-19th century. China is an exciting venture. It is one of the greatest human experiments of our time—like the Green Revolution, or the manned space flight program. The economic development of a vast continent and even vaster populace is an enormous challenge. The Asian tigers are likely to have been just the precursor of development in Asia. While we have not seen the first dominant innovation—like a Chinese iPod, wonder drug, or Windows operating system, there is no reason to think that contributions from China like these will not be forthcoming and perhaps soon. China has assimilated contract high-end manufacturing, it is moving into contract design. It would be a mistake to bet against China's earning a very respectable place in the forefront of innovation. The only question is when.

IV

APPENDIXES

Appendix A

Agenda

20-22 SEPTEMBER 2006 COOPERATIVE SYMPOSIUM
"INNOVATIVE FLANDERS:
INNOVATION POLICIES FOR THE 21ST CENTURY"

LEUVEN, FLANDERS

Day 1: Wednesday, 20 September 2006

12:00 U.S. Speakers' Lunch

13:30 Registration

14:00 Welcoming Remarks & Overview of the Programme
dr. Peter Spyns, Department of Economy, Science and Innovation, The Flemish Government

14:15 Remarks by the U.S. Delegation
dr. Bill Spencer, SEMATECH (retired)

14:30 ***Session I: Perspectives on the Flemish Innovation System***
Moderator: Charles W. Wessner, PhD, U.S. National Research Council

The Flemish Innovation System and Its Components
dr. Peter Spyns, Department of Economy, Science and Innovation, The Flemish Government

Implementation and Monitoring of the Flemish Innovation System
Eric Sleeckx, Flemish Institute for the Promotion of Innovation through Science & Technology (IWT)

Discussant
Prof. dr. Luc Soete, University of Maastricht, Netherlands, & UN Univ-MERIT

15:30 **Coffee Break**

16:30 ***Session II: Perspectives on the U.S. Innovation System***
Moderator: Prof. dr. Luc Soete United Nations University, Director of MERIT

Challenges and Current Developments in the U.S. Innovation System
Mary Good, University of Arkansas at Little Rock

Global Competition, Corporate Policy, and National Interest
Mark B. Myers, Xerox Corporation (retired)

18:00 ***Ministerial Address***
Fientje Moerman, Vice Minister-President of the Flemish Government, Minister for Economy, Enterprise, Science, Innovation, and Foreign Trade

18:30 ***End of day 1***

Day 2: Thursday, 21 September 2006

09:00 ***Session III: Cooperative Research and Global Competition in Semiconductors***
Moderator: dr. Peter Spyns, Department of Economy, Science and Innovation, The Flemish Government

Current Trends—A U.S. Industry Perspective
George Scalise, Semiconductor Industry Association

Chinese Innovation & Development Strategies
Alan Wm. Wolff, Dewey Ballantine, L.L.P.

The IMEC Concept and Contribution
Anton De Proft, IMEC

The SEMATECH Contribution
Kenneth Flamm, University of Texas at Austin

The IMEC Experience—A Strategic Partner Industry Perspective
Allen Bowling, Texas Instruments

10:40 Coffee Break

11:10 ***Session IV: Innovation through Knowledge Diffusion***
Moderator: Mark B. Myers, Xerox Corporation (retired)

The Leuven Experience as Centre of an Innovative Regional Hot Spot
Prof. dr. ir. Koenraad Debackere, K.U.Leuven

An Industry Perspective: The Case of the Chemical Industry
dr. Erwin Annys, Federation of the Belgian Chemical Industries and Life Sciences (formerly Fedichem, now Essenscia)

University-Based Entrepreneurship in the United States
Prof. dr. Paul Ducheyne, University of Pennsylvania

12:25 Lunch

13:30 ***Session V: Meeting the Early-stage Finance Challenge***
Moderator: Prof. dr. Luc Soete, University of Maastricht, Netherlands, and UN Univ-MERIT

Texas Innovation Initiatives
Pike Powers, Fulbright & Jaworski, L.L.P.

Overview of TxAN: A New Model for Research Collaboration
Randal K. Goodall, SEMATECH

From Research to University Spin-off: Experience in Belgium
Prof. dr. Bruno de Vuyst, Vesalius College, Vrije Universiteit Brussel (VUB) and Lawfort Brussels

Commercializing University Research: The Role of SBIR
Charles W. Wessner, PhD, U.S. National Research Council

Flemish Innovation Fund: Goals, Mechanisms, and Results
Rudy Aernoudt, Department of Economy, Science, and Innovation

Impact of the Policy Measures on High-tech Innovative Start-ups
Prof. dr. B. Clarysse, Blerick Leuven Gent Management School, Ghent University

Concept and Evaluation of the Advanced Technology Program
Marc Stanley, National Institute of Standards and Technology

The Challenge of Collecting Good Evaluation Data
Prof. dr. Bart Van Looy, Flemish Research Center for Science, Technology, and Innovation Research (SOOS)

17:10 ***End of day 2***

Day 3: Friday, 22 September 2006

10:00 ***Session VI: Flemish Strategic Research Centres***
Moderator: Peter Spyns, Flemish Authorities, Department of Economy, Science and Innovation, Technology and Innovation Division

Flanders Interuniversity Institute for Biotechnology (VIB)
dr. Lieve Ongena, Flanders Interuniversity Institute for Biotechnology (VIB)

Flemish Institute for Technological Research (VITO)
Dirk Fransaer, Flemish Institute for Technological Research (VITO)

10:50 **Coffee Break**

Interdisciplinary Institute for Broadband Technology (IBBT)
Wim De Waele, Interdisciplinary Institute for Broadband Technology (IBBT)

General Closure

11:30 ***Tour of IMEC Facility***

12:00 **Lunch**

13:30 ***End of Day 3***

Appendix B

Biographies of Speakers*

RUDY AERNOUDT

Rudy Aernoudt has studied economy and philosophy at the University of Leuven and holds a master's degree in European economics from the College of Europe. After a career as corporate manager in the banking sector, he became principal administrator at the European Commission dealing with enterprise policy, in particular the financing of European enterprises. He was special adviser to the Belgian President of the Industry Council during the Belgian Presidency of the European Union before becoming deputy head of cabinet to the Walloon minister of economics, dealing with research, entrepreneurship, and financing. Afterwards he became director of cabinet to the federal minister of economics, energy, external trade, and scientific policy. From 2004 onwards, he became chief of staff (head of cabinet) to the Flemish vice-president and minister of economy, enterprises, science, and international trade. From September 1, 2006, to September 16, 2007, he was secretary general of the Department of Economics, Science, and Innovation of the Flemish Government.

Rudy Aernoudt is also professor in corporate finance at the business schools of Brussels and Ghent and the European University of Nancy. He is the author of 20 books and over 150 articles on the topics of finance and politics. He is a member of the editorial boards of *Venture Capital*, an international journal of entrepreneurial finance, *Financial Management*, and *Entrepreneurial and Business*

*As of September 2006.

Angel Financing. He is a speaker at numerous conferences organized by different institutions such as the United Nations, Organization for Economic Co-operation and Development, University of Antwerp Management School, European Private Equity and Venture Capital Association, Vlerick Management School, Sorbonne II, the European Parliament, AECM (Association Européenne du Cautionnement Mutuel), European Business Angel Network, Wall Street Journal Entrepreneurship Summit, Wirtschafts symposium, European Business School, and the Belgian Venture Capital Academy. He was the responsible organiser of the yearly European symposium on financing in Louvain-Ia-Neuve, gathering over 3,000 participants, and is a member of the boards of different enterprises.

ERWIN ANNYS

Erwin Annys obtained his Ph.D. in chemistry at the University of Gent. After 16 years in the chemical industry as well in a multinational as in a medium-sized company, he started working at Fedichem, the Federation of the Belgian Chemical Industries, where he is responsible for product and innovation policy with activities on the Flemish, Belgian, and European levels. He is a member of the European Chemical Industry Council Innovation Planning Group, SusChem, and the board of the Belgian Normalisation, and he is the author of a book on REACH, the coming European legislation on chemicals.

R. ALLEN BOWLING

Allen Bowling is TI Fellow and manager of external research for Texas Instruments' Silicon Technology Development (SiTD) group in Dallas, Texas. Allen coordinates SiTD's involvement in external research in the area of silicon process/materials/devices technology, including SEMATECH, International SEMATECH Manufacturing Initiative (ISMI), Semiconductor Research Corporation (SRC), IMEC, and various university research programs.

Dr. Bowling currently represents Texas Instruments (TI) on the following advisory groups: (1) SEMATECH Executive Steering Council, (2) ISMI Executive Advisory Council, (3), SRC Executive Technical Advisory Board, (4) SRC Nanomanufacturing Science Area Coordinating Committee, (5) IMEC Operations Review Meeting group, and (6) University of Arizona ESH Research Center, with which he serves as chair of the Executive Advisory Council.

Dr. Bowling completed his Ph.D. in chemistry in 1979 at the University of Tennessee. He held an Alexander von Humboldt research fellowship from 1979 to 1980 at the University of Frankfurt, Germany. He joined TI in 1980 as a process control engineer in the TI Dallas MOS-2 wafer fab. From 1982 to 1987, he was a member of the Technical Staff in the TI Central Research Labs, Materials Science Laboratory. His research focused on particulate contamination control, wafer cleaning, and process control in semiconductor process technology. He and

Graydon Larrabee did the research leading up to the TI Microelectronics Manufacturing Science and Technology (MMST) program on single-wafer processing for flexible manufacturing, administered by the Office of Naval Research and the U.S. Air Force. In 1987, he was elected TI Senior Member of the Technical Staff. From 1987 to 1997, he was a manager for process development in TI's Semiconductor Process and Device Center (SPDC). From 1997 to 2001, he was a manager for process development in TI SiTD. Dr. Bowling moved to his role as SiTD Manager of External Research in November 2001. In 2002, Allen was elected TI Fellow, one of 67 TI Fellows currently within TI.

KOENRAAD DEBACKERE

Koenraad Debackere has been with K.U.Leuven since 1995. He obtained his master degrees in electromechanical engineering (1984) and management sciences (1985) and his Ph.D. degree in management with an ICM-fellowship at the University of Gent (1990), after stays as an ICM-fellow (1988-1989) and an ICRMOT research assistant (1990) at MIT Sloan School of Management. He was a Fulbright-Hays postdoctoral fellow at MIT from 1991 to 1992. In 1992, he became an assistant professor at Erasmus University Rotterdam and an NFWO-postdoctoral researcher in 1993. In 1995, he became professor at K.U.Leuven where he teaches technology and innovation management.

In 1993, 1995, and 1997, Koenraad Debackere won Best Research Paper Awards from the American Academy of Management and the Decision Sciences Institute. His research has focused on the area of technology and innovation management and policy, the development of indicators for measuring the linkage between science and technology, the design and use of bibliometric indicators for science policy purposes and the role of entrepreneurial universities in economic development. He has published over 150 articles and book chapters in this field. Publications have appeared in *Research Policy*, *Scientometrics*, *Journal of Product Innovation Management*, *R&D Management*, *Technovation*, *Technological Forecasting and Social Change*, *Journal of the American Society for Information Science*, *Small Business Economics*, *Proceedings of the American Academy of Management*, *International Journal of Management Reviews*, *EMBO Reports*, *Research Evaluation*, *The Economics of Innovation and New Technology*, *International Journal of Technology Management*, *Journal of High Technology*, *Management Research*, *Creativity and Innovation Management*, *The Journal of Technology Transfer*, and *Research Technology Management*.

Koenraad Debackere has also been an invited professor in the area of innovation management in various academic programs (Manchester Business School, Insead, Milano Politechnico, Tilburg University, and Chalmers University). He has been a promotor and recipient of various research grants by The Institute for the Promotion of Innovation by Science and Technology in Flanders (IWT), Research Foundation—Flanders (FWO), DWTC (Belgium), and the European

Commission. He is promoter of "Steunpunt O&O Statistieken" of the Flemish government at K.U.Leuven.

Koenraad Debackere is also actively engaged in technology transfer activities as managing director of K.U.Leuven research and development and chairman of Gemma Frisius Fonds (the venture fund) of the K.U.Leuven. He is the co-founder and chairman of Leuven, Inc., the innovation network of Leuven high-tech entrepreneurs. He is a board member of IWT-Vlaanderen, the Flemish government agency that supports science and technology development in Flemish industry. Since 2003 he has been member of the board of K.U.Leuven, and since 2005 he has been the general manager of K.U.Leuven.

PAUL DUCHEYNE

Paul Ducheyne is professor of bioengineering, professor of orthopedic surgery research, and professor of biomaterials in dentistry at the University of Pennsylvania in Philadelphia. He also is the director of the Center for Bioactive Materials and Tissue Engineering at this university.

Paul Ducheyne, a native of Belgium, went through his secondary school years in the latin-mathematics-humanities as *primus perpetuus*. Upon graduation in 1967, he attended the Catholic University of Leuven, where he obtained an engineering degree in materials science and engineering in 1972 and a Ph.D. in 1976.

Paul Ducheyne is generally considered a leader in biomaterials and tissue engineering. He has organized a number of symposia and meetings, such as the Fourth European Conference on Biomaterials (1983), the Engineering Foundation Conference on Bioceramics (1986), which led to the New York Academy of Sciences publication, *Bioceramics, Material Characteristics Versus In Vivo Behavior*, and the Sixth International Symposium on Ceramics in Medicine (1993). He has lectured around the world and served on the editorial board of ten scientific journals in the biomaterials, bioceramics, bioengineering, tissue engineering, orthopedics, and dental fields. He has authored more than 280 papers and chapters in a variety of international journals and books and he has edited ten books. When last quantified (December 2004), his number of citations was an average of 160 per year over his career with a maximum of 580 in 1992. Paul Ducheyne has been secretary of the European Society for Biomaterials (1980-1983) and president of the Society for Biomaterials (USA) (1994-1995). During his tenure as president, the National Institutes of Health (NIH) moved forward with the organization of the workshop on "Biomaterials and Medical Implant Science—Future Directions." He is also past president of the International Society for Ceramics in Medicine (1993) and currently is a member of its board of directors. He has been the first Nanyang Visiting Professor at the Nanyang Institute of Technology, Singapore (1999-2000), and is a member of the International Advisory Board to the EEC program on Tissue Engineering (2004-present) involving 14 European countries. He is a fellow of the American Association for the Advancement of Science, the American Institute for Medical and Biological Engineering, and the International Association of Biomaterials Societies.

Paul Ducheyne obtained his Ph.D. on a thesis entitled "Metallic Orthopaedic Implants with a Porous Coating" in 1976. With fellowships from the NIH (International Postdoctoral Fellowship) and the Belgian American Educational Foundation (Honorary Fellowship), he performed postdoctoral research at the University of Florida (the Laboratory of Bioceramics and the Department of Orthopaedic Surgery). It was at the University of Florida that he began his research on bioactive ceramics. He first presented his seminal work on the use of hydroxyapatite coatings on porous materials for enhancing fixation to bone in 1977. This concept is now used worldwide in orthopedic clinical care.

Paul Ducheyne started his career in Europe, specifically at his alma mater, the Catholic University of Leuven. He created the now highly successful post-graduate program in bioengineering in the School of Engineering and Applied Sciences. Among the four co-founders was the previous president of this University, Professor A. Oosterlinck. In those initial years, he was also chairman-founder of the chapter on biomedical engineering of the Belgian Engineering Society (Flemish section) and director of Meditek, the Flemish government body created to promote Academia to Industry Technology Transfer in the area of biomedical engineering.

Dr. Ducheyne joined the University of Pennsylvania in 1983. From 1991 to 1994 he was chair of the university's graduate group in bioengineering, the then largest Ph.D. program in bioengineering in the United States. As graduate group chair, he focused on enhancing the overall quality of the graduate student body and the competitiveness of the university's fellowships in bioengineering. He also initiated the programmatic changes that led to the formulation of tracks in bioengineering.

Paul Ducheyne directs a research program with research personnel that have come from all parts of the world. Support for his work has come from federal agencies (National Science Foundation [NSF]), NIH, Veterans Administration, NASA, and DARPA) and industry. His laboratory is currently funded by NIH grants, including a Bioengineering Research Partnership grant, and grants from DARPA and the Pennsylvania Nanotechnology Institute. His proposals have frequently drawn outstanding reviews, as can be witnessed by the 0.2 percentile on an NIH proposal, the first ranking in NSF competition, or the size of a 1980 grant from a leading U.S. corporation to his laboratory when still in Europe. His program has active collaborations with several overseas institutions. Notable is his collaboration with Professor Gutmanas at the Technion, Israel, and the repeated support from the Binational Science Foundation, Israel.

Paul Ducheyne's work led to a number of firsts and to concepts now used in industry. A number of patents followed from his work and, over the years, led to many collaborations with leading medical device companies. He also assisted corporations in the review of legal proceedings and their technical merit. As such he was expert witness in many high-visibility cases, most notably the porous coating patent litigation of the second half of the 1980s, the temporo-mandibular joint (TMJ) litigation which instigated the Biomaterials Availability Act, the breast implant controversy, and the pedicle screw dossier.

Since 1990, all technologies coming from his lab were withheld from licensing to existing companies. These patents formed the wealth of technology at the basis of Orthovita (NASDAQ: VITA), a corporation founded in 1992 with offices in the Philadelphia area and operations in the United States and Belgium. Paul Ducheyne attracted Orthovita's top-level management, and he served as chairman of the board until 1999 and director until 2003.

From 2001 to 2002, Paul Ducheyne was on leave from academia and directed the start-up phase of Gentis, Inc., Philadelphia. Gentis, Inc. is now well under way to becoming the leader in treating low back pain resulting from degenerating intervertebral discs. This condition is highly prevalent already in a middle-aged population, and leads to the largest number of sick days in our Western societies. A minimally invasive procedure has been designed that will enable resolution of pain, quick regain of function and, in general, postponement of disease progression and major spine fusion surgery.

KENNETH S. FLAMM

Kenneth S. Flamm is professor and Dean Rusk Chair in International Affairs at the Lyndon Baines Johnson School (LBJ School) of Public Affairs at the University of Texas at Austin. He is a 1973 honors graduate of Stanford University and received a Ph.D. in economics from MIT in 1979. From 1993 to 1995, Dr. Flamm served as Principal Deputy Assistant Secretary of Defense for Economic Security and special assistant to the Deputy Secretary of Defense for Dual Use Technology Policy. Prior to and after his service at the Defense Department, he spent 11 years as a senior fellow in the Foreign Policy Studies Program at the Brookings Institution. Dr. Flamm has been a professor of economics at the Instituto Tecnológico A. de México in Mexico City, the University of Massachusetts, and George Washington University.

Dr. Flamm currently directs the LBJ School's Technology and Public Policy Program and directs externally funded research projects on "Internet Use in Developing and Industrializing Countries," "The Economics of Fair Use," and "Determinants of Internet Use in U.S. Households," and has recently initiated a new project on "Exploring the Digital Divide: Regional Differences in Patterns of Internet Use in the U.S." He continues to work with the semiconductor industry research consortium, International SEMATECH, and is building a return-on-investment-based prototype to add economic logic to SEMATECH's industry investment model. He also is a member of the National Research Council's Committees on the Future of Supercomputing, Measuring and Sustaining the

New Economy, and Comparative Innovation Policy. He has served as member and chair of the NATO Science Committee's Panel for Science and Technology Policy and Organization, and as a member of the Federal Networking Council Advisory Committee, the Organization for Economic Co-operation and Development Expert Working Party on High Performance Computers and Communications, and various advisory committees and study groups of the National Science Foundation, the Council on Foreign Relations, the Defense Science Board, and the U.S. Congress' Office of Technology Assessment, and as a consultant to government agencies, international organizations, and private corporations.

Dr. Flamm is the author of numerous articles and books on the economic impacts of technological innovation in a variety of high-technology industries. Among the latter are *Mismanaged Trade? Strategic Policy and the Semiconductor Industry* (1996), *Changing the Rules: Technological Change, International Competition, and Regulation in Communications* (ed., with Robert Crandell, 1989), *Creating the Computer* (1988), and *Targeting the Computer* (1987). Recent work by Dr. Flamm has focused on measurement of the economic impact of the semiconductor industry on the U.S. economy, analyzing the economic determinants of Internet use by households, and assessing the economic impacts of Internet use in key applications.

MARY L. GOOD

Mary L. Good is the Donaghey University Professor at the University of Arkansas at Little Rock, and serves as dean for the College of Information Science and Systems Engineering. She is managing member for the Fund for Arkansas' Future, LLC (an investment fund for startup and early-stage companies), past president of the American Association for the Advancement of Science, past president of the American Chemical Society, and an elected member of the National Academy of Engineering. She presently serves on the Boards of BiogenIdec, Inc., and Acxiom, Inc.

Previously Dr. Good served a 4-year term as the under secretary for technology for the Technology Administration in the Department of Commerce, a presidentially appointed, Senate confirmed, position. In addition, she chaired the National Science and Technology Council's Committee on Technological Innovation (NSTC/CTI), and served on the NSTC Committee on National Security. Previously she has served as the senior vice president for Technology for Allied Signal and as the Boyd Professor of Chemistry and Materials Science at Louisiana State University.

Dr. Good was appointed to the National Science Board by President Carter in 1980 and by President Reagan in 1986. She was the chair of that board from 1988 to 1991, when she received an appointment by President Bush to be a member of the President's Council of Advisors on Science and Technology.

Mary Good has received many awards, including the National Science Foundation's Distinguished Public Service Award, the American Institute of Chemists' Gold Medal, the Priestly Medal from the American Chemical Society, and the Vannevar Bush Award from the National Science Board, among others.

Dr. Good received her bachelor's degree in chemistry from the University of Central Arkansas and her M.S. and Ph.D. degrees in inorganic chemistry from the University of Arkansas at Fayetteville.

RANDAL GOODALL

Randal Goodall is the director of external programs at SEMATECH in Austin, Texas, a consortium of the world's leading computer chip makers. His career has combined scientific research, technology development, systems engineering, and new business and funding development.

Dr. Goodall received his bachelor of science in physics from Caltech (1977) and his master's degree (1979) and Ph.D. degree (1984) from the University of Oregon in experimental solid-state physics, studying the Quantum Hall Effect in semiconductor devices. After graduation he transferred skills developed in laboratory automation to lead the product development efforts of an advanced software applications startup as director of engineering.

In 1987 Dr. Goodall joined ADE in Boston, a leading producer of advanced measurement systems for the semiconductor industry. He formed the Systems Technology Group to identify and develop next generation micro- and nanoscale measurement technologies, system architectures, and computational applications.

In early 1994, Dr. Goodall joined SEMATECH as a senior member of the Technical Staff in the Silicon Materials Group working on the world's earliest 300mm wafer materials and equipment development efforts. In late 1995, Dr. Goodall was one of six members of the startup team for the International 300mm Initiative (I300I), leading the Enabling Technologies division, including the silicon wafer, metrology, standards, and productivity programs. In 1998, the I300I programs merged with International SEMATECH, and in 2000, he was named associate director of a new $13 million Manufacturing Methods and Productivity division, focusing on productivity for existing and future fabs and equipment. He spear-headed efforts in global technology collaboration and standardization, including invention of a novel new computational model of the entire industry to support R&D "what if" analysis. He participated in the formation of a new manufacturing R&D subsidiary to offer these programs more broadly.

Beginning in 2001 on special assignment to the Office of the Chief Executive, Dr. Goodall worked to secure $200 million of leveraged funding for the Albany Extreme Ultraviolet patterning program. In 2002, he co-developed the Texas Technology Initiative (TTI), a comprehensive technology-based economic development platform, which enabled a proactive funding response by the State of Texas. He worked with the Texas governor and other state and local officials to pass $295 million 2003 legislation which enabled funding for SEMATECH and university programs through a new Advanced Materials Research Center (AMRC), spanning semiconductor, biotechnology, nanotechnology, MEMS, and advanced energy. Acting as technology coordinator, Dr. Goodall managed the 2004 formation of AMRC, which continues to perform coordinated industry-university R&D, and the new Advanced Processing and Prototyping Center (AP2C), funded by DARPA, which provides R&D infrastructure for emerging technologies. As the first director of the newly formed SEMATECH External Programs office, he provided leadership in 2005 for the TTI, the State Strategy on Advanced Technology, and the definition and passage of the $200 million Texas Emerging Technology Fund legislation.

Dr. Goodall has been engaged in the most dynamic components of SEMATECH's evolution for more than 12 years. He continues to work with local, state, and national government efforts to drive technology innovation and economic development, and he partners with technology leaders, university administrators and researchers, and state officials to develop mechanisms for co-leveraging the semiconductor infrastructure of SEMATECH/ATDF and the nanofabrication needs of emerging technologies. He is a leader in the new Texas Alliance for Nanoelectronics.

Dr. Goodall has published dozens of scientific and technical papers on solid-state physics, silicon wafer technology, R&D collaboration, industry technology transitions, including 300mm wafers, and productivity modeling.

FIENTJE MOERMAN

Fientje Moerman is the vice minister-president of the Flemish Government and Flemish minister for Economy, Enterprise, Science, Innovation and Foreign Trade, a post she has held since 2004. She received her law degree from the University of Ghent (honors), specializing in tax and economic law, and completed a master of law degree at Harvard Law School, where she was a Fulbright Scholar. In 1983, she completed the New York State Bar exam.

From 1982 to 1984, Fientje Moerman was a lawyer at the firm of Cleary, Gottlieb, Steen & Hamilton in New York and Brussels. She then took a position as deputy editor for economics and finance at the *Standaard*. From 1985 to 1989, she served as spokesperson for the Liberals and Democrats at the European Parliament, and in 1989 she took a position as adviser to the president of France, Valéry Giscard d'Estaing, a post she held until 1991. She then served as chief adviser for institutional reforms and relations with Israel and the Gulf States from 1991 to 1995.

Fientje Moerman is a member of the Party Executive Committee of the Flemish Liberals (PVV, now VLD), in which capacity she has served from 1991 to 1993 and again from 1997 to the present. She is a member of the Flemish Liberal Students Association (LVSV) and of the Council of European Liberals

(ELDR) and is chairperson of "Politiek Konvent," the umbrella organization of political groupings at the University of Ghent. She served as municipal councilor for the City of Ghent from 1988 to 1995 and vice mayor for education for that city from 1995 to 1999. Minister Moermann served as a member of parliament from 1999 to 2003, holding the post of secretary-general of the VLD. In 2003, she became minister for Economy, Energy, Foreign Trade and Science Policy, and in 2004 she became vice minister-president of the Flemish Government and Flemish Minister for Economy, Enterprise, Science, Innovation and Foreign Trade.

MARK B. MYERS

Mark B. Myers is a consultant in the fields of R&D management, emerging technology trends, entrepreneurial startups, and national innovation policies. He served on the National Resarch Council's Board on Science, Technology, and Economic Policy (STEP) from 1994 to 2005, where he was the co-chair of the STEP study, A Patent System for the 21st Century. Dr. Myers recently completed an appointment as the Walter C. Bladstrom Visiting Executive Professor in Entrepreneurial Management at the University of Pennsylvania's Wharton School (2004-2005) and Visiting Executive Professor of Management (2002–2005), where his research interests included identifying emerging markets and technologies to enable growth in new and existing companies with special emphases on technology identification and selection, product development, and technology competencies. He retired from the Xerox Corporation in 2000, after a 37-year career in its R&D laboratories. He was the Xerox senior vice president in charge of corporate research, advanced development, systems architecture and corporate engineering from 1992 to 2000. His responsibilities included the corporate research centers in Palo Alto, California; Webster, NewYork; Toronto Canada; Cambridge, UK; and Grenoble, France. During this period, he was also a member of the senior management committee in charge of the strategic direction setting of the company. He holds a bachelor's degree from Earlham College and a doctorate in materials science from the Pennsylvania State University.

PIKE POWERS

A partner since 1978, Pike Powers is partner-in-charge of Fulbright & Jaworski L.L.P.'s Austin office. Mr. Powers was executive assistant to Governor Mark White in 1983 and from 1972 to 1979 represented Jefferson County in the Texas House of Representatives. He has extensive experience in handling complex legal and political issues before state courts and federal courts, as well as federal and state agencies.

Mr. Powers has been a member of the board of directors of the State Bar of Texas and has held various posts as well in the American Bar Association and in the Texas and American Bar Foundations. He is a former chairman of the Board of the Austin Chamber of Commerce. Mr. Powers is a member of the Maritime Law Association of the United States, the Federation of Insurance and Corporate Counsel, and the National Association of Railroad Trial Counsel.

Mr. Powers was named as a "Texas Super Lawyer" in general litigation law in the November 2003 issue of *Texas Monthly*.

Mr. Powers received a B.A. in 1962 from Lamar University and a J.D. in 1965 from The University of Texas. He was admitted in 1965 to practice law in Texas.

GEORGE M. SCALISE

George M. Scalise is president of the Semiconductor Industry Association (SIA) where he directs a staff focused on International Trade & Government Affairs, Workforce, Technology, Environmental-Safety & Health, and Communications.

Scalise came to the SIA from Apple Computer, where he served as executive vice president of operations. Prior to that, he held executive management positions at National Semiconductor, Maxtor Corporation, Advanced Micro Devices, Fairchild Semiconductor, and Motorola Semiconductor.

A graduate of Purdue University with a bachelor of science degree in mechanical engineering, Mr. Scalise is a highly respected technology industry spokesperson and carries a special interest and expertise in technology, international trade, and competition issues. He was a founding member of the Semiconductor Research Corporation, an industry-funded organization that provides resources for pre-competitive semiconductor research at American universities.

Mr. Scalise currently serves on President George W. Bush's Council of Advisors on Science and Technology as well as numerous boards, including Cadence Design Systems, Intermolecular, and iSuppli Corporation. He has also served on the boards at the Federal Reserve Bank of San Francisco (where he was chairman and was chairman of the executive committee of the Conference of Chairmen of the Federal Reserve System), SEMATECH, Semiconductor Research Corporation, the Bay Area Economic Forum, and Dubai Silicon Oasis, and was a member of the Council on Foreign Relations Economic Task Force on Japan. He participates on advisory committees at the College of Engineering at Purdue University and is a member of the California Council on Science and Technology Fellows Program. He served on the advisory committees at the Leavey School of Business at Santa Clara University and the School of Engineering at the University of Southern California. He was named a Distinguished Engineering Alumnus of Purdue University in 2002. He also chaired the Secretary of Energy Advisory Board, U.S. Department of Energy.

ERIC SLEECKX

Eric Sleeckx started his career in 1982 as mechanical engineer at the University Hospital in Leuven. He was responsible for the development of custom-made implants. In 1986 he moved to Philips Industrial activities where he became project leader for the development of the Compact Disc Video player. From 1988 to 1998 he was responsible for the research team, Product Innovation, at WTCM (research centre for the technological industry). Finally he joined The Institute for the Promotion of Innovation by Science and Technology in Flanders where he has been coordinating a variety of innovation support programs and the CIN (cooperative innovation networks) and since autumn 2005 has coordinated the Monitoring and Analysis unit.

WILLIAM J. SPENCER

William J. Spencer was named chairman emeritus of the SEMATECH Board in November 2000 after serving as chairman of the SEMATECH and International SEMATECH Boards since July 1996. He came to SEMATECH in October 1990 as president and chief executive officer. He continued to serve as president until January 1997 and CEO until November 1997. During this time, SEMATECH became totally privately funded and expanded to include non-U.S. members. Many gave SEMATECH part of the credit for the U.S. semiconductor turnaround in the 1990s.

Dr. Spencer has held key research positions at Xerox Corporation, Bell Laboratories, and Sandia National Laboratories. Before joining SEMATECH in October 1990, he was group vice president and senior technical officer at Xerox Corporation in Stamford, Connecticut, from 1986 to 1990. He established new research centers in Europe and developed a plan for Xerox retaining ownership in spinout companies from research. Prior to joining the Xerox Palo Alto Research Center (PARC) as manager of the Integrated Circuit Laboratory in 1981 and as the Center Manager of PARC in 1982 to 1986, Dr. Spencer served as director of Systems Development from 1978 to 1981 at Sandia National Laboratories in Livermore, and director of Microelectronics at Sandia National Laboratories in Albuquerque from 1973 to 1978, where he developed a silicon processing facility for Department of Energy needs. He began his career in 1959 at Bell Laboratories.

Dr. Spencer received the Regents Meritorious Service Medal from the University of New Mexico in 1981; the C. B. Sawyer Award for contribution to "The Theory and Development of Piezoelectric Devices" in 1972; and a Citation for Achievement from William Jewell College in 1969, where he also received a doctor of science degree in 1990. He is a member of the National Academy of Engineering, a Fellow of the Institute of Electrical and Electronics Engineers, and serves on numerous advisory groups and boards, including the Committee on Comparative Innovation Policy at the National Research Council. He was the Regents Professor at the University of California in the spring of 1998. He has been a visiting professor at the University of California at Berkeley School of Engineering and the Haas School of Business since the fall of 1998. He is a research professor of medicine at the University of New Mexico.

PETER SPYNS

Peter Spyns has studied at the Katholieke Universiteit Leuven (M.A. in romance philology, and M.Sc. and Ph.D. in computer science). He has worked as a researcher on various European Union (EU) projects in the field of medical language and knowledge processing at the university hospital in Leuven and, later on, in Ghent (each time at the division of Medical Informatics).

After obtaining his Ph.D. in 1996, Dr. Spyns joined the speech and language technology company Lernout & Hauspie Speech Products (L&H), once a worldwide leader in its domain. He was active as a principal linguistic engineer in the area of spoken dialogue systems. For one year he was seconded to the joint venture between Intel Corporation and L&H, during which he assisted the chief technology officer with assessment of speech and language processing tools and information extraction and retrieval technologies.

After the collapse of L&H, Dr. Spyns returned in 2002 to academia (Vrije Universiteit Brussel) where he became a postdoc senior researcher in the field of ontology engineering and the semantic web.

Since beginning 2006, Dr. Spyns has been a senior researcher at the Technology & Innovation unit of the Department of Economy, Science, and Innovation of the Flemish Government. His main activities concern the innovation policy in Flanders. In that respect, he currently participates in the EU ERA-net project, VISION, on innovation policy preparation and represents Flanders in some national and international committees (Organization for Economic Co-operation and Development Working Group on Innovation and Technology Policy, GSO Trendchart on Innovation, and VRWB-CTB).

In addition, Dr. Spyns coordinates the joint research program between Flanders and the Netherlands on speech and language technology for Dutch (STEVIN). For these tasks, he is seconded to the Dutch Language Union. In his spare time, he still performs some scientific research work on ontologies and the semantic web for the Vrije Universiteit Brussel.

MARC G. STANLEY

Marc G. Stanley has been the director of the Advanced Technology Program (ATP) at the National Institute of Standards and Technology (NIST) since June 2003. He also serves as a U.S. Governor on the Israel-U.S. Binational Industrial Research and Development (BIRD) Foundation Board of Governors and as the American Director on the Trilateral Industrial Development (TRIDE) Executive Committee.

Mr. Stanley served as the acting director of ATP from 2001 to 2003 and as the associate director for the program from 1993 to 2001.

Before coming to NIST, Mr. Stanley was the associate deputy secretary of the U.S. Department of Commerce (DoC) by presidential appointment. He served as counselor to the NIST Director, as a consultant to DoC's Technology Administration, and as assistant secretary for congressional and intergovernmental affairs at DoC.

Mr. Stanley earned a B.A. from George Washington University and a bachelor of law degree from the University of Baltimore.

JAMES TURNER

James Turner has served on the professional staff of the Committee on Science in the U.S. House of Representatives for over 25 years. He currently serves as the full committee chief democratic counsel where he works across the board on the committee's legislative agenda.

For the 10 years prior to the Republican takeover of Congress, Mr. Turner was the committee's senior staff member for technology policy including 4 years as technology subcommittee staff director. He also served as a subcommittee legal counsel. During the late 1970s and early 1980s, he worked on the committee's Republican staff as minority energy counsel. His legislative interests include the international competitiveness of U.S. industry, environmental and energy research and development, trade and technology policy, intellectual property, standards, and technology transfer. Mr. Turner's work has been recognized over the years through awards presented by the American Mathematical Society, American National Standards Institute, American Society for Mechanical Engineering, ASTM International, Federal Laboratory Consortium for Technology Transfer, Federal Patent Law Association, National Institute for Building Science, Semiconductor Industry Association, Technology Transfer Society, U.S. Metric Association, and the Virginia Engineering Foundation.

Mr. Turner also spent 3 years working for Wheelabrator-Frye, 2 years for Congressman Gary Myers, 2 years for the State of Connecticut, and shorter periods with the National Aeronautics and Space Administration and the Federal Aviation Administration. He holds degrees from Georgetown and Yale Universities and from Westminster College and attended the Senior Managers in Government Program at Harvard University.

Mr. Turner currently serves on the Boards of Trustees of University of Virginia's (UVA's) School of Engineering, the Accelerating Innovation Foundation, and ASTM International; on the Advisory Boards for Carnegie Mellon's Heinz School; The Science, Technology and Society Program at UVA, and the journal *Innovations*, published by MIT Press; and on the Vestry of St. Columba's Episcopal Church in Washington, D.C. Mr. Turner also is the Washington coordinator for the MIT and UVA joint Washington Summer Intern Program.

BART VAN LOOY

Bart Van Looy is professor at K.U.Leuven in the field of innovation and organization, within the Department of Managerial Economics, Strategy and Innovation (Faculty of Economics and Applied Economics). His current research focuses on organizing innovation (company level) and regional innovation systems: entrepreneurial universities and science-technology interactions are focal points of attention in this respect. Bart Van Looy has published on these topics in journals like *Research Policy,* the *Journal of Product and Innovation Management, R&D Management, Scientometrics,* and the *Journal of Technology Transfer*. He is also first editor of an international text book on services management (FT/Prentice Hall; 1998/2003; Japanese edition: 2005; Chinese edition, 2006).

Bart Van Looy has been at K.U.Leuven since 1998. He holds master's degrees in psychology and applied economics. He obtained his Ph.D. in organizational psychology at K.U.Leuven in 2000. His research is focused in the area of technology and innovation management (project/company level), the role of entrepreneurial universities in economic development, science-technology interactions in conjunction with technological development trajectories, and knowledge intensive ventures. Bart Van Looy is also actively engaged in the European Institute of Advanced Studies in Management (EIASM). Currently he is responsible at INCENTIM for several research projects carried out in close collaboration with K.U.Leuven R&D, DWTC (Belgium), the European Commission, and Steunpunt O&O Statistieken (Flemish Region), where he is also engaged as senior scientist responsible for the patent database infrastructure and the related service and research activities. In October 2005 Bart Van Looy was appointed as professor at K.U.Leuven within the field of innovation and organization teaching the following courses: Introduction to Management (bachelor/master level); Innovation Management (master level) and Innovation and Technology Management (master after master level). He is also teaching the course Managerial Research Methods within the MBA program of the Flanders Business School (Antwerp) where he recently became appointed as scientific coordinator. Before joining K.U.Leuven, he worked as a consultant within the field of HRM, organizational behavior and organizational design (1990-1996), and as a researcher at the Vlerick Management School, University of Ghent (1996-1998).

CHARLES W. WESSNER

Charles W. Wessner is a policy advisor recognized nationally and internationally for his expertise on innovation policy, including public-private partnerships, entrepreneurship, early-stage financing for new firms, and the special needs and benefits of high-technology industry. He testifies to the U.S. Congress and major national commissions, advises agencies of the U.S. government and international organizations, and lectures at major universities in the United States and abroad. Reflecting the strong global interest in innovation, he is frequently asked to address issues of shared policy interest with foreign governments, universities, and research institutes, often briefing government ministers and senior officials.

Dr. Wessner's work addresses the linkages between science-based economic growth, entrepreneurship, new technology development, university-industry clusters, regional development, small firm finance, and public-private partnerships. His program at the National Academies also addresses policy issues associated with international technology cooperation, investment, and trade in high-technology industries. Currently, he directs a series of studies centered on government measures to encourage entrepreneurship and to support the development of new technologies. Foremost among these is a congressionally mandated study of the Small Business Innovation Research (SBIR) program, reviewing the operation and achievements of this $2 billion award program for small companies and startups. A major review of the technology drivers of the New Economy and its sustained productivity growth is nearing completion. He is also directing a major new study on best practice in global innovation programs, entitled *Comparative Innovation Policy: Best Practice for the 21st Century.*

ALAN WM. WOLFF

Alan Wm. Wolff is Managing Partner of Dewey Ballantine's Washington, D.C. office and chairs Dewey Ballantine's International Trade Group.

Founded by Alan Wolff in 1979, the International Trade Group consists of 24 lawyers, 11 economists and analysts, and 12 research assistants, and is known for taking an interdisciplinary approach to trade issues drawing on in-depth resources in law, policy, and economic and factual research. The Group is involved on behalf of clients in major initiatives to open international markets for both goods and services as well as representing clients in a broad range of trade litigation before government agencies, courts, NAFTA bi-national panels, and dispute settlement in the World Trade Organization. Its activities include dealing with matters involving government regulations and tariffs, antidumping and subsidy cases, competition policy, intellectual property rights, and investment regulations. The Group also regularly conducts investigations and represents clients in enforcement proceedings regarding a broad range of regulatory requirements including export regulations, sanctions, boycotts, foreign corrupt practices, money laundering, and various statutory compliance audits.

Alan Wolff also serves as chairman of the Advisory Board of The Institute for Trade & Commercial Diplomacy, and is currently a member of the National Research Council's Board on Science, Technology, and Economic Policy, the U.S. Department of State's Advisory Committee on International Economic Policy, the Advisory Committee of the Institute for International Economics, the Advisory Board of the Economic Strategy Institute, the Board of Trustees of the United States Council for International Business, the Board of Advisors of the American

Health and Education Foundation, the Council on Foreign Relations, the American Society of International Law, and the American Bar Association.

Mr. Wolff served as United States Deputy Special Representative for Trade Negotiations (1977-1979), holding the rank of ambassador; he was General Counsel of the agency from 1974 to 1977. From 1968 to 1973, Mr. Wolff was an attorney dealing with international monetary, trade, and development issues at the Treasury Department. Mr. Wolff has co-authored books and published numerous papers on trade and U.S. trade law, many of which are listed on the International Trade Group's web site (<*http://www.dbtrade.com*>).

Mr. Wolff received an LL.B. from Columbia University in 1966 and an A.B. from Harvard College in 1963. He is a member of the bar in Massachusetts, New York, the District of Columbia, the U.S. Court of International Trade, the Court of Appeals for the Federal Circuit, and the Supreme Court of the United States.

Appendix C

Bibliography

Acs, Zoltan and David Audretsch. 1990. *Innovation and Small Firms.* Cambridge, MA: The MIT Press.

Aerts, Kris and Dirk Czarnitzki. 2005. "Using Innovation Survey Data to Evaluate R&D Policy: The Case of Flanders." Catholic University of Leuven, Department of Applied Economics and Steunpunt O&O Statistieken.

Aerts, Kris and Dirk Czarnitski. 2006. "The Impact of Public R&D Funding in Flanders." *IWT Studies*, 54.

Aghion, Phillipe, Robin Burgess, Stephen Redding, and Fabrizio Zilibotti. 2003. "The Unequal Effects of Liberalization: Theory and Evidence from India." Center for Economic Policy Research.

Ahluwalia, Montek Singh. 2001. "State Level Reforms Under Economic Reforms in India." Stanford University Working Paper No. 96, March.

Alic, John A., Lewis M. Branscomb, Harvey Brooks, Ashton B. Carter, and Gerald L. Epstein. 1992. *Beyond Spin-off: Military and Commercial Technologies in a Changing World*. Boston, MA: Harvard Business School Press.

Amsden, Alice H. 2001. *The Rise of "the Rest": Challenges to the West from Late-industrializing Economies*. Oxford, UK: Oxford University Press.

Amsden, Alice H. and Wan-wen Chu. 2003. *Beyond Late Development: Taiwan's Upgrading Policies*. Cambridge, MA: The MIT Press.

Amsden, Alice H., Ted Tschang, and Akira Goto. 2001. "Do Foreign Companies Conduct R&D in Developing Countries?" Tokyo, Japan: ADB Institute.

Archibugi, Danielle, Jeremy Howells, and Jonathan Michie, eds. 1999. *Innovation Policy and the Global Economy*. Cambridge, UK: Cambridge University Press.

Asheim, Bjorn T. et al., eds. 2003. *Regional Innovation Policy for Small-medium Enterprises*. Cheltenham, UK and Northampton, MA: Edward Elgar.

Athreye, Suma S. 2000. "Technology Policy and Innovation: The Role of Competition Between Firms." In Pedro Conceicao et al., eds. *Science, Technology, and Innovation Policy: Opportunities and Challenges for the Knowledge Economy*. Westport, CT and London, UK: Quorum Books.

Audretsch, David B., ed. 1998. *Industrial Policy and Competitive Advantage*, Volumes 1 and 2. Cheltenham, UK: Edward Elgar.

Audretsch, David B. 2006. *The Entrepreneurial Society.* Oxford, UK: Oxford University Press.
Audretsch, David B., Heike Grimm, and Charles W. Wessner. 2005. *Local Heroes in the Global Village: Globalization and the New Entrepreneurship Policies.* New York: Springer.
Audretsch, D. B., B. Bozeman, K. L. Combs, M. P. Feldman, A. N. Link, D. S. Siegel, P. Stephan, G. Tassey, and C. Wessner. 2002. "The Economics of Science and Technology." *Journal of Technology Transfer* 27:155-203.
Baker, Stephen. 2005. "New York's Big Hopes for Nano." *Businessweek.* February 4.
Baldwin, John Russel, and Peter Hanel. 2003. *Innovation and Knowledge Creation in an Open Economy: Canadian Industry and International Implications.* Cambridge, UK: Cambridge University Press.
Balzat, Markus and Andreas Pyka. 2006. "Mapping National Innovation Systems in the OECD Area." *International Journal of Technology and Globalisation* 2(1-2):158-176.
Bartzokas, Anthony and Morris Teubal. 2002. "The Political Economy of Innovation Policy Implementation in Developing Countries." *Economics of Innovation and New Technology* 11(4-5).
Bhidé, Amar. 2006. "Venturesome Consumption, Innovation and Globalization." Paper presented at the Centre on Capitalism & Society and CESifo Venice Summer Institute 2006, "Perspectives on the Performance of the Continent's Economies," 21-22 July 2006. Held at Venice International University. San Servolo, Italy.
Biegelbauer, Peter S. and Susana Borras, eds. 2003. *Innovation Policies in Europe and the U.S.: The New Agenda.* Aldershot, UK: Ashgate.
Birch, David. 1981. "Who Creates Jobs?" *The Public Interest*, 65:3-14.
Blomström, Magnus, Ari Kokko, and Fredrik Sjöholm. 2002. "Growth & Innovation Policies for a Knowledge Economy: Experiences from Finland, Sweden, & Singapore." EIJS Working Paper, Series No. 156.
Bloomberg News. 2006. "The Next Green Revolution." August 21.
Booz Allen Hamilton. 2005. "Booz Allen Hamilton Global Innovation 1000: Money Isn't Everything." *Strategy + Business,* Issue 41.
Borras, Susana. 2003. *The Innovation Policy of the European Union: From Government to Governance.* Cheltenham, UK: Edward Elgar.
Borrus, Michael and Jay Stowsky. 2000. "Technology Policy and Economic Growth." In Charles Edquist and Maureen McKelvey, eds. *Systems of Innovation: Growth, Competitiveness and Employment, Vol. 2.* Cheltenham, UK and Northampton, MA: Edward Elgar.
Branscomb, Lewis M. and Philip E. Auerswald. 2002. *Between Invention and Innovation: An Analysis of Funding for Early-Stage Technology Development.* NIST GCR 02–841. Gathersburg, MD: National Institute of Standards and Technology. November.
Braudel, Fernand. 1973. *Capitalism and Material Life 1400-1800.* London: Harper Colophon Books.
Buchanan, James M. 1987. "An Economic Theory of Clubs." In *Economics: Between a Predictive Science and Moral Philosophy.* College Station, TX: Texas A&M University Press, 1987.
Bush, Nathan. 2005. "Chinese Competition Policy, It Takes More than a Law." *China Business Review.* May-June.
Capron, Henri and Michele Cincera. 2006. Strengths and Weaknesses of the Flemish Innovation System: An External Viewpoint. Brussles: IWT.
Caracostas, Paraskevas and Ugur Muldur. 2001. "The Emergence of the New European Union Research and Innovation Policy," in P. Laredo and P. Mustar, eds. *Research and Innovation Policies in the New Global Economy: An International Comparative Analysis.* Cheltenham, UK: Edward Elgar.
Chand, Satish and Kunal Sen. 2002. "Trade Liberalization and Productivity Growth: Evidence from Indian Manufacturing." *Review of Development Economics* 6, February.
Chesbrough, Henry. 2003. *Open Innovation: The New Imperative for Creating and Profiting from Technology.* Cambridge, MA: Harvard Business School Press. April.

China Trade Extra. 2005. *China Agrees to Delay Software Procurement Rule While Talking with U.S.* July 11.

Chinese Ministry of Finance. 2006. *Opinions of the Ministry of Finance on Implementing Government Procurement Policies That Encourage Indigenous Innovation.* Cai Ku [2006] No. 47, June 13.

Cimoli, Mario and Marina della Giusta. 2000. "The Nature of Technological Change and its Main Implications on National and Local Systems of Innovation." IIASA Interim Report IR-98-029.

Cincera, Michele. 2006. R&D Activities of Flemish Companies in the Private Sector: An Analysis for the Period 1998-2002. Brussels: IWT.

Cincera, Michele. 2006. Comparison of Regional Approaches to Foster Innovation in the European Union: The Case of Flanders. Brussels: IWT.

Coburn, Christopher and Dan Berglund. 1995. *Partnerships: A Compendium of State and Federal Cooperative Programs.* Columbus, OH: Battelle Press.

Combs, Kathryn L. and Albert N. Link. 2003. Innovation Policy in Search of an Economic Paradigm: The Case of Research Partnerships in the United States. *Technology Analysis & Strategic Management* 15(2).

Council on Competitiveness. 2005. *Innovate America: Thriving in a World of Challenge and Change.* Washington, D.C.: Council on Competitiveness.

Crafts, N. F. R. 1995. "The Golden Age of Economic Growth in Western Europe, 1950-1973." *Economic History Review* 3:429-447.

Czarnitzki, Dirk and Niall O'Byrnes. 2007. "Innovation and the Impact on Productivity in Flanders." *Tijdschrift voor Economie en Management* 52(2).

Dahlman, Carl J. 2005. *India and the Knowledge Economy: Leveraging Strengths and Opportunities.* Washington, D.C.: World Bank Publications.

Dahlman, Carl J. and Jean Eric Aubert. 2001. *China and the Knowledge Economy: Seizing the 21st Century.* Washington, D.C.: The World Bank.

Dahlman, Carl and Anuja Utz. 2005. *India and the Knowledge Economy: Leveraging Strengths and Opportunities.* Washington, D.C.: The World Bank.

Daneke, Gregory A. 1998. "Beyond Schumpeter: Non-linear Economics and the Evolution of the U.S. Innovation System." *Journal of Socio-economics* 27(1):97-117.

Das, Gurcharan. 2006. "The India Model." *Foreign Affairs* 85(4).

Davidsson, Per. 1996. "Methodological Concerns in the Estimation of Job Creation in Different Firm Size Classes." Working Paper, Jönköping International Business School.

Davis, Steven, John Haltiwanger, and Scott Schuh. 1993. "Small Business and Job Creation: Dissecting the Myth and Reassessing the Facts." Working Paper No. 4492. Cambridge, MA: National Bureau of Economic Research.

Debackere, Koenraad and Wolfgang Glänzel. 2004. "Using a Bibliometric Approach to Support Research Policy Making: The Case of the Flemish BOF-key." *Scientometrics* 59(2).

Debackere, Koenraad and Reinhilde Veugelers. 2005. "The Role of Academic Technology Transfer Organizations in Improving Industry Science Links." *Research Policy* 34(3):321-342.

de Jonquieres, Guy. 2004. "China and India Cannot Fill the World's Skills Gap." *Financial Times.* July 12.

de Jonquieres, Guy. 2004. "To Innovate, China Needs More than Standards." *Financial Times.* July 12.

de Jonquieres, Guy. 2006. "China's Curious Marriage of Convenience." *Financial Times.* July 19.

De la Mothe and Gilles Paquet. 1998. "National Innovation Systems, 'Real Economies' and Instituted Processes." *Small Business Economics* 11:101-111.

Doloreux, David. 2004. "Regional Innovation Systems in Canada: A Comparative Study." *Regional Studies* 38(5):479-492.

Domain-b.com. 2005. "Nasscom-McKinsey: India to Face Skilled Workers' Shortage by Next Decade." Dec. 12. Available online at <*http://www.domain-b.com/organisation/nasscom/20051217_shortage.html*>.

Dries, Ilse, Peer van Humbeek, and Jan Larosse. 2005. "Linking Innovation Policy and Sustainable Development in Flanders." Paris: OECD.

Eaton, Jonathan, Eva Gutierrez, and Samuel Kortum. 1998. "European Technology Policy." NBER Working Papers 6827.

Edler, J. and S. Kuhlmann. 2005. "Towards One System? The European Research Area Initiative, the Integration of Research Systems and the Changing Leeway of National Policies." *Technikfolgenabschätzung: Theorie und Praxis*. 1(4):59-68.

EE Times. 2006. "Chinese Province Pays to Get 300-mm Wafer Fab." June 28.

Eickelpasch, Alexander and Michael Fritsch. 2005. "Contests for Cooperation: A New Approach in German Innovation Policy." *Research Policy* 34:1269-1282.

Electronic News. 2006. "SMIC Gets $3B Nod from Chain's Wuhan Government." May 22.

Endquist, Charles, ed. 1997. *Systems of Innovation: Technologies, Institutions, and Organizations*. London, UK: Pinter.

EOS Gallup Europe. 2004. *Entrepreneurship*. Flash Eurobarometer 146. January. Accessed at <*http://ec.europa.eu/enterprise/enterprise_policy/survey/eurobarometer146_en.pdf*>.

European Commission. 2003. "Innovation in Candidate Countries: Strengthening Industrial Performance." May.

European Commission. 2003. "Investing in Research: An Action Plan for Europe 2003." Brussels: EC.

European Commission. 2003. *3rd S&T Indicators Report*. Luxembourg: Office for Official Publications of the European Communities.

Fangerberg, Jan. 2002. *Technology, Growth, and Competitiveness: Selected Essays*. Cheltenham, UK and Northampton, MA: Edward Elgar.

Federal Register Notice. 2004. "2004 WTO Dispute Settlement Proceeding Regarding China: Value-Added Tax on Integrated Circuits." April 21.

Feldman, Maryann and Albert N. Link. 2001. "Innovation Policy in the Knowledge-based Economy." *Economics of Science, Technology and Innovation Volume 23*. Boston, MA: Kluwer Academic Press.

Feldman, Maryann P., Albert N. Link, and Donald S. Siegel. 2002. *The Economics of Science and Technology: An Overview of Initiatives to Foster Innovation, Entrepreneurship, and Economic Growth*. Boston, MA: Kluwer Academic Press.

The Financial Times. 2006. "India Needs Big Infrastructure Drive." February 23.

Flamm, Kenneth. 2003. "SEMATECH Revisited: Assessing Consortium Impacts on Semiconductor Industry R&D." In National Research Council. *Securing the Future: Regional and National Programs to Support the Semiconductor Industry*. Charles W. Wessner, ed. Washington, D.C.: The National Academies Press.

Fonfria, Antonio, Carlos Diaz de la Guardia, and Isabel Alvarez. 2002. "The Role of Technology and Competitiveness Policies: A Technology Gap Approach." *Journal of Interdisciplinary Economics* 13(1-2-3):223-241.

Foray, Dominique and Patrick Llerena. 1996. "Information Structure and Coordination in Technology Policy: A Theoretical Model and Two Case Studies." *Journal of Evolutionary Economics* 6(2).

Fortune. 2006. "America's Most Admired Companies 2006."

Friedman, Thomas. 2005. *The World Is Flat: A Brief History of the 21st Century*. New York: W. H. Freeman.

Furman, Jeffrey L., Michael E. Porter, and Scott Stern. 2002. "The Determinants of National Innovative Capacity." *Research Policy* 31:899-933.

George, Gerard and Ganesh N. Prabhu. 2003. "Developmental Financial Institutions as Technology Policy Instruments: Implications for Innovation and Entrepreneurship in Emerging Economies." *Research Policy* 32(1):89-108.

Government of the People's Republic of China. 2004. *Anticompetitive Practices of Multinational Companies and Countermeasures*. Administration of Industry and Commerce, Office of Anti-monopoly, Fair Trade Bureau, State Administration of Industry and Commerce, May.

Government of the People's Republic of China, Office of Anti-monopoly, Fair Trade Bureau, State Administration of Industry and Commerce. 2004. "Anticompetitive Practices of Multinational Companies in China and Countermeasures." *Journal of Administration of Industry and Commerce*. May.

Government of the People's Republic of China. 2005. *Article 10: Exemptions of Monopoly Agreements*. Anti-Monopoly Law of the People's Republic of China. Revised July 27.

Government of the People's Republic of China, Ministry of Information Industry. 2006. "Outline of the 11th Five-Year Plan and Medium-and-Long-Term Plan for 2020 for Science and Technology Development in the Information Industry." Xin Bu Ke, No. 309, August 29.

Government of the People's Republic of China, National Development and Reform Commission. 2006. The 11th Five-Year Plan. March 19. Available online at *<http://english.gov.cn/2006-07/26/content_346731.htm>*.

Grande, Edgar. 2001. "The Erosion of State Capacity and European Innovation Policy: A Comparison of German and EU Information Technology Policies." *Research Policy* 30(6):905-921.

Grindley, Peter, David Mowery, and Brian Silverman. 1994. "SEMATECH and Collaborative Research: Lessons in the Design of High Technology Consortia." *Journal of Policy Analysis and Management* 13(4):723-758.

Guo Ban Han No. 30. 2006. Letter from the General Office of the State Council on Approving the Formulation of the Rules for Implementation of the Several Supporting Policies for Implementation of the Outline of the National Medium and Long-term Plan for Development of Science and Technology. *Gazette of the State Council*. Issue No. 17, Serial No. 1196, June 20.

Hall, Bronwyn H. 2002. "The Assessment: Technology Policy." *Oxford Review of Economic Policy* 18(1):1-9.

Hu, Zhijian. 2006. "IPR Policies In China: Challenges and Directions." Presentation at *Industrial Innovation in China*. Levin Institute Conference. July 24-26.

Huang, Yasheng and Tarun Khanna. 2003. "Can India Overtake China?" *Foreign Policy* July–August.

Hughes, Kent. 2005. *Building the Next American Century: The Past and Future of American Economic Competitiveness*. Washington, D.C.: Woodrow Wilson Center Press. Chapter 14.

Hughes, Kent H. 2005. "Facing the Global Competitiveness Challenge." *Issues in Science and Technology*. XXI(4):72–78.

Hulsink, W., H. Bouwman, and T. Elfring. 2007. "Silicon Valley in the Polder? Entrepreneurial Dynamics, Virtuous Clusters and Vicious Firms in the Netherlands and Flanders." ERIM Report Series Research in Management (ERS-2007-048-ORG).

Inside U.S.-China Trade. 2006. *Industry Worried China Backing out of Commitment to Join GPA*. September 27.

International Herald Tribune. 2006. "Start-Ups Explore Abroad for IPOs." December 25.

Jaffe, Adam B., Josh Lerner, and Scott Stern, eds. 2003. *Innovation Policy and the Economy: Volume 3*. Cambridge, MA: The MIT Press.

Jasanoff, Sheila, ed. 1997. *Comparative Science and Technology Policy*. Elgar Reference Collection. International Library of Comparative Pubic Policy, Volume 5. Cheltenham, UK and Lyme, NH: Edward Elgar.

Jorgenson, Dale and Kevin Stiroh. 2002. "Raising the Speed Limit: Economic Growth in the Information Age." In National Research Council. *Measuring and Sustaining the New Economy*. Dale Jorgenson and Charles Wessner, eds. Washington, D.C.: The National Academies Press.

Joy, William. 2000. "Why the Future Does Not Need Us." *Wired* 8.04. April.

Kapur, Devesh. 2003. "Indian Diaspora as a Strategic Asset." *Economic and Political Weekly* 38(5):445-448.

Kapur, Surinder. 2006. "Nurture New Technology and Innovation, Stay Competitive." *The Financial Express*. November 16.

The Kauffman Foundation and the Information Technology and Innovation Foundation. 2007. *The 2007 State New Economy Index*. Available online at <*http://www.kauffman.org/items.cfm?itemID=766*>.

Kim, Yong-June. 2006. "A Korean Perspective on China's Innovation System." Presentation at *Industrial Innovation in China*. Levin Institute Conference. July 24-26.

Koizumi, Kei. 2007. "Historical Trends in Federal R&D." In *AAAS Report XXXII: Research and Development FY2008*. AAAS Publication Number 07-1A. Washington, D.C.: American Association for the Advancement of Science.

Koschatzky, Knut. 2003. "The Regionalization of Innovation Policy: New Options for Regional Change?" In G. Fuchs and Phil Shapira, eds. *Rethinking Regional Innovation: Path Dependency or Regional Breakthrough?* London, UK: Kluwer.

Kuhlmann, Stephan and Jakob Edler. 2003. "Scenarios of Technology and Innovation Policies in Europe: Investigating Future Governance—Group of 3" *Technological Forecasting & Social Change* 70.

Kumar, Deepak. 1995. *Science and the Raj: 1857–1905*. New Delhi and New York: Oxford University Press.

Lall, Sanjaya. 2002. "Linking FDI and Technology Development for Capacity Building and Strategic Competitiveness." *Transnational Corporations* 11(3):39-88.

Laredo, Philippe and Philippe Mustar, eds. 2001. *Research and Innovation Policies in the New Global Economy: An International Perspective*. Cheltenham, UK: Edward Elgar.

Larosse, J. 2004. "Towards a 'Third Generation' Innovation Policy in Flanders: Policy Profile of the Flemish Innovation." Brussels: IWT.

Lembke, Johan. *Competition for Technological Leadership: EU Policy for High Technology.* Cheltenham, UK and Northampton, MA: Edward Elgar, 2002.

Lemola, Tarmo. 2002. "Convergence of National Science and Technology Policies: The Case of Finland." *Research Policy* 31(8-9):1481-1490.

Lerner, Joshua. 1999. "Public Venture Capital." In National Research Council. *The Small Business Innovation Program: Challenges and Opportunities*. Charles W. Wessner, ed. Washington, D.C.: National Academy Press.

Leslie, Stuart and Robert Kargon. 2006. "Exporting MIT." *Osiris* 21:110-130.

Lewis, James A. 2005. *Waiting for Sputnik: Basic Research and Strategic Competition*. Washington, D.C.: Center for Strategic and International Studies.

Li, Jianjun. 2006. "The Development and Opening of Tianjin Binhai: New Area & China's Biotechnical Innovations." Presentation at *Industrial Innovation in China*. Levin Institute Conference. July 24-26.

Lin, Otto. 1998. "Science and Technology Policy and its Influence on the Economic Development of Taiwan." In Henry S. Rowen, ed., *Behind East Asian Growth: The Political and Social Foundations of Prosperity*. London, UK and New York, NY: Routledge.

Long, Guogiang. 2005. "China's Policies on FDI: Review and Evaluation." In Theodore H. Moran, Edward M. Graham, and Magnus Blomström, eds., *Does Foreign Direct Investment Promote Development?* Washington, D.C.: Institute for International Economics.

Lynn, Leonard. 2006. "Collaborative Advantage and China's Evolving Position in the Global Technology System." Presentation at *Industrial Innovation in China*. Levin Institute Conference. July 24-26.

Macher, Jeffrey, David Mowery, and David Hodges. 1999. "Semiconductors." In National Research Council. *U.S. Industry in 2000: Studies in Competitive Performance*. David C. Mowery, ed. Washington, D.C.: National Academy Press.

Maddison, Angus and Donald Johnston. 2001. *The World Economy: A Millennial Perspective*. Paris, France: Organization for Economic Cooperation and Development.

Mani, Sunil. 2004. "Government, Innovation and Technology Policy: An International Comparative Analysis." *International Journal of Technology and Globalization* 1(1).

McKibben, William. 2003. *Enough: Staying Human in an Engineered Age*. New York: Henry Holt & Co.
Meyer-Krahmer, Frieder. 2001. "The German Innovation System." Pp. 205-252 in P. Larédo and P. Mustar, eds. *Research and Innovation Policies in the New Global Economy: An International Comparative Analysis*. Cheltenham, UK: Edward Elgar.
Meyer-Krahmer, Frieder. 2001. "Industrial Innovation and Sustainability—Conflicts and Coherence." Pp. 177-195 in Daniele Archibugi and Bengt–Ake Lundvall, eds. *The Globalizing Learning Economy.* New York: Oxford University Press.
Mitra, Raja. 2006. "India's Potential as a Global R&D Power." In Magnus Karlsson, ed. *The Internationalization of Corporate R&D*. Östersund: Swedish Institute for Growth Policy Studies.
Mody, Ashok and Carl Dahlman. 1992. "Performance and Potential of Information Technology: An International Perspective." *World Development* 20(12):1703-1719.
Moore, Gordon. 2003. "The SEMATECH Contribution." In National Research Council. *Securing the Future: Regional and National Programs to Support the Semiconductor Industry*. Charles W. Wessner, ed. Washington, D.C.: The National Academies Press.
Murali, Kanta. 2003. "The IIT Story: Issues and Concerns." *Frontline* 20(3).
Murphy, L. M. and P. L. Edwards. 2003. *Bridging the Valley of Death: Transitioning from Public to Private Sector Financing*. Golden, CO: National Renewable Energy Laboratory, May.
Mustar, Phillipe and Phillipe Laredo. 2002. "Innovation and Research Policy in France (1980–2000) or The Disappearance of the Colbertist State." *Research Policy* 31:55-72.
National Academy of Engineering. 2004. *The Engineer of 2020: Visions of Engineering in the New Century*. Washington, D.C.: The National Academies Press.
National Academy of Sciences/National Academy of Engineering/Insitute of Medicine. 2007. *Rising Above the Gathering Storm: Energizing and Employing America for a Brighter Economic Future*. Washington, D.C.: The National Academies Press.
National Manufacturing Competitiveness Council. 2006. "The National Strategy for Manufacturing." New Delhi, March.
National Research Council. 1996. Conflict and Cooperation in National Competition for High-technology Industry. Washington, D.C.: National Academy Press.
National Research Council. 1999. *The Advanced Technology Program: Challenges and Opportunities.* Charles W. Wessner, ed. Washington, D.C.: National Academy Press.
National Research Council. 1999. *Funding a Revolution: Government Support for Computing Research*. Washington, D.C.: National Academy Press.
National Research Council. 1999. *Industry-Laboratory Partnerships: A Review of the Sandia Science and Technology Park Initiative.* Charles W. Wessner, ed. Washington, D.C.: National Academy Press.
National Research Council. 1999. *New Vistas in Transatlantic Science and Technology Cooperation.* Charles W. Wessner, ed. Washington, D.C.: National Academy Press.
National Research Council. 1999. *The Small Business Innovation Research Program: Challenges and Opportunities.* Charles W. Wessner, ed. Washington, D.C.: National Academy Press.
National Research Council. 2000. *The Small Business Innovation Research Program: A Review of the Department of Defense Fast Track Initiative.* Charles W. Wessner, ed. Washington, D.C.: National Academy Press.
National Research Council. 2000. *U.S. Industry in 2000: Studies in Competitive Performance*. David C. Mowery, ed. Washington, D.C.: National Academy Press.
National Research Council. 2001. *The Advanced Technology Program: Assessing Outcomes.* Charles W. Wessner, ed. Washington, D.C.: National Academy Press.
National Research Council. 2001. *Building a Workforce for the Information Economy.* Washington, D.C.: National Academy Press.
National Research Council. 2001. *Capitalizing on New Needs and New Opportunities: Government-Industry Partnerships in Biotechnology and Information Technologies*. Charles W. Wessner, ed. Washington, D.C.: National Academy Press.

National Research Council. 2001. *A Review of the New Initiatives at the NASA Ames Research Center.* Charles W. Wessner, ed. Washington, D.C.: National Academy Press.

National Research Council. 2001. *Trends in Federal Support of Research and Graduate Education.* Stephen A. Merrill, ed. Washington, D.C.: National Academy Press.

National Research Council. 2002. *Government-Industry Partnerships for the Development of New Technologies: Summary Report.* Charles W. Wessner, ed. Washington, D.C.: The National Academies Press.

National Research Council. 2003. *Government-Industry Partnerships for the Development of New Technologies: Summary Report.* Charles W. Wessner, ed. Washington, D.C.: The National Academies Press.

National Research Council. 2003. *Securing the Future: Regional and National Programs to Support the Semiconductor Industry*. Charles W. Wessner, ed. Washington, D.C.: The National Academies Press.

National Research Council. 2004. *The Small Business Innovation Research Program: Program Diversity and Assessment Challenges.* Charles W. Wessner, ed. Washington, D.C.: The National Academies Press.

National Research Council. 2005. *Getting Up to Speed: The Future of Superconducting*, Susan L. Graham, Marc Snir, and Cynthia A. Patterson, eds. Washington, D.C.: The National Academies Press.

National Research Council. 2005. *Policy Implications of International Graduate Students and Postdoctoral Scholars in the United States*. Washington, D.C.: The National Academies Press.

National Research Council. 2007. *Enhancing Productivity Growth in the Information Age: Measuring and Sustaining the New Economy*. Dale W. Jorgenson and Charles W. Wessner, eds. Washington, D.C.: The National Academies Press.

National Research Council. 2007. *Innovation Policies for the 21st Century*. Charles W. Wessner, ed. Washington, D.C.: The National Academies Press.

National Research Council. 2007. *India's Changing Innovation System*. Charles W. Wessner and Sujai J. Shivakumar, eds. Washington, D.C.: The National Academies Press.

National Research Council. 2007. *SBIR and the Phase III Challenge of Commercialization.* Charles W. Wessner, ed. Washington, D.C.: The National Academies Press.

National Research Council. 2008. *An Assessment of the Small Business Innovation Research Program.* Charles W. Wessner, ed. Washington, D.C.: The National Academies Press.

National Science Board. 2004. *Science and Engineering Indicators 2004.* Arlington, VA: National Science Foundation.

National Science Foundation. 2004. *Science and Engineering Doctorate Awards: 2003*, NSF 05-300, Arlington, VA: National Science Foundation.

Needham, Joseph. 1954-1986. *Science and Civilization in China* (five volumes). Cambridge: Cambridge University Press.

Nelson, Richard R. and Katherine Nelson. 2002. "Technology, Institutions, and Innovation Systems." *Research Policy* 31:265-272.

Nelson, Richard R. and Nathan Rosenberg. 1993. "Technical Innovation and National Systems." In *National Innovation Systems: A Comparative Analysis*. Richard R. Nelson, ed. Oxford, UK: Oxford University Press.

Organisation for Economic Co-operation and Development. 1999. *Boosting Innovation: The Cluster Approach*. Paris, France: Organisation for Economic Co-operation and Development.

Organisation for Economic Co-operation and Development. 1999. *Managing National Innovation Systems*. Paris, France and Washington, D.C.: Organisation for Economic Co-operation and Development.

Organisation for Economic Co-operation and Development. 2001. *Social Sciences and Innovation*. Washington, D.C.: Organisation for Economic Co-operation and Development.

Organisation for Economic Co-operation and Development. 2004. "Summary Report: Micro-policies for Growth and Productivity." DSTI/IND(2004)7. Paris, France: Organization for Economic Cooperation and Development. October.

Organisation for Economic Co-operation and Development. 2006. *Main Science and Technology Indicators*. Paris: Organisation for Economic Co-operation and Development.

Orszag, Peter and Thomas Kane. 2003. "Funding Restrictions at Public Universities: Effects and Policy Implications." *Brookings Institution Working Paper*, September.

Oughton, Christine. 1997. "Competitiveness in the 1990s." *The Economic Journal* 107(444):1486-1503.

Oughton, Christine, Mikel Landabaso, and Kevin Morgan. 2002. "The Regional Innovation Paradox: Innovation Policy and Industrial Policy." *The Journal of Technology Transfer* 27(1).

Patel, P. and K. Pavitt. 1994. "National Innovation Systems: Why They are Important and How They Might Be Compared?" *Economic Change and Industrial Innovation*.

Posen, Adam S. 2001. "Japan." In Benn Steil, David G. Victor, and Richard R. Nelson, eds. *Technological Innovation and Economic Performance*. Princeton, NJ: Princeton University Press.

President's Council of Advisors on Science and Technology. 2004. "Sustaining the Nation's Innovation System: Report on Information Technology Manufacturing and Competitiveness." Washington, D.C.: Executive Office of the President. January.

PricewaterhouseCoopers. 2006. "China's Impact on the Semiconductor Industry: 2005 Update." PricewaterhouseCoopers.

Procassini, Andrew. 1995. *Competitors in Alliance: Industry Associations, Global Rivalries, and Business-Government Relations*. New York: Greenwood Publishing.

Raduchel, William. 2006. "The End of Stovepiping." In National Research Council. *The Telecommunications Challenge: Changing Technologies and Evolving Policies*, Charles W. Wessner, ed., Washington, D.C.: The National Academies Press.

Rai, Saritha. 2006. "India Becoming a Crucial Cog in the Machine at I.B.M." *The New York Times*. June 5.

Reid, T. R. 2004. *The United States of Europe: The New Superpower and the End of American Supremacy*. Penguin Press.

Reuters. 2006. "China Sees No Quick End to Economic Boom." February 21.

Rodrik, Dani and Arvind Subramanian. 2004. "From 'Hindu Growth' to Productivity Surge: The Mystery of the Indian Growth Transition." NBER Working Paper 10376.

Romanainen, Jari. 2001. "The Cluster Approach in Finnish Technology Policy." Pp. 377-388 in Edward M. Bergman, Pim den Hertog, and David Charles, eds., *Innovative Clusters: Drivers of National Innovation Systems*. OECD Proceedings. Washington, D.C.: Organisation for Economic Co-operation and Development.

Ruttan, Vernon. 2002. *Technology, Growth and Development: An Induced Innovation Perspective*. Oxford, UK: Oxford University Press.

Rutten, Roel and Frans Boekema. 2005. "Innovation, Policy and Economic Growth: Theory and Cases." *European Planning Studies* 13(8).

Schaffer, Teresita. 2002. "Building a New Partnership with India." *Washington Quarterly* 25(2):31-44.

Scherer, F. M. 2001. "U.S. Government Programs to Advance Technology." *Revue d'economie industrielle* 0(94):69-88.

Semiconductor Fabtech. 2006. "Chinese Government Builds 300mm Fab." June 28.

Semiconductor Manufacturing International Corp. 2006. Form 20-F for Period Ending December 31, 2006.

Semiconductor Manufacturing International. 2006. "Q1 2006 Earnings Conference Call." April 27.

Shang, Yong. 2006. "Innovation: New National Strategy of China." Presentation at *Industrial Innovation in China*. Levin Institute Conference. July 24-26.

Shanghai Municipal Government. 2004. *Notice of the Shanghai Municipal Government Regarding Distributing the Outline of Shanghai's Intellectual Property Strategy (2004-2010)*, September 14.

Sheehan, Jerry and Andrew Wyckoff. 2003. "Targeting R&D: Economic and Policy Implications of Increasing R&D Spending." DSTI/DOC(2003)8. Paris, France: Organization for Economic Cooperation and Development.

Shenzhen Daily. 2005. "Nation May Introduce Antimonopoly Law." December 30.

Shin, Roy W. 1997. "Interactions of Science and Technology Policies in Creating a Competitive Industry: Korea's Electronics Industry." *Global Economic Review* 26(4):3-19.

Singh, Manmohan. 2006. Remarks by the Chairman of the Planning Commission on the release of the 11th Plan Approach Paper. Accessed at <*http://pmindia.nic.in/speech/content.asp?id=431*> October 18.

Smits, Ruud and Stefan Kuhlmann. 2004. "The Rise of Systemic Instruments in Innovation Policy." *International Journal of Foresight and Innovation Policy* 1(1/2).

Soete, Luc G. and Bastiann J. ter Weel. 1999. "Innovation, Knowledge Creation and Technology Policy: The Case of the Netherlands." *De Economist* 147(3). September.

Spencer, William and T. E. Seidel. 2004. "International Technology Roadmaps: The U.S. Semiconductor Experience." In National Research Council. *Productivity and Cyclicality in Semiconductors: Trends, Implications, and Questions*. Dale W. Jorgenson and Charles W. Wessner, eds. Washington, D.C.: The National Academies Press.

Stanford University. 1999. *Inventions, Patents and Licensing: Research Policy Handbook*. Document 5.1. July 15.

State Council of the Peoples' Republic of China. 2006. *Outline of the National Medium- and Long-term Program on Scientific and Technological Development (2006-2020)*. February 9.

Talele, Chitram J. 2003. "Science and Technology Policy in Germany, India and Pakistan." *Indian Journal of Economics and Business* 2(1):87-100.

Tansley, A. G. 1939. "British Ecology During the Past Quarter Century: The Plant Community and the Ecosystem." *The Journal of Ecology* 27(2):513-530.

Tassey, Gregory. 2004. "Policy Issues for R&D Investment in a Knowledge-based Economy." *Journal of Technology Transfer* 29:153-185.

Teubal, Morris. 2002. "What Is the Systems Perspective to Innovation and Technology Policy and How Can We Apply It to Developing and Newly Industrialized Economies?" *Journal of Evolutionary Economics* 12(1-2).

TMCNET. 2006. "Chip Giant Quickens Steps of Expansion." June 28.

U.S.–China Joint Commission on Commerce and Trade (JCCT). 2005. *Outcomes on Major U.S. Trade Concerns*. Washington, D.C.: The Office of the United States Trade Representative.

U.S. Department of Energy. 2006. Press Release. "Department Requests $4.1 Billion Investment as Part of the American Competitiveness Initiative: Funding to Support Basic Scientific Research." February 2.

U.S. General Accounting Office. 2002. *Export Controls: Rapid Advances in China's Semiconductor Industry Underscore need for Fundamental U.S. Policy Review*. GAO-020620. Washington, D.C.: U.S. General Accounting Office. April.

Van Looy, Bart, K. Debackere, and T. Magerman. 2005. *Assessing Academic Patent Activity: The Case of Flanders*. Leuven: SOOS.

Van Looy, Bart, Marina Ranga, Julie Callaert, Koenraad Debackere, and Edwin Zimmermann. 2004. "Combining Entrepreneurial and Scientific Performance in Academia: Towards a Compounded and Reciprocal Matthew-effect?" *Research Policy* 33(3):425-441.

Vervliet, Greta. 2006. *Science, Technology, and Innovation*. Ministry of Flanders, Science and Innovation Administration.

Veugelers, Reinhilde, Jan Larosse, Michele Cincera, Donald Carchon, and Roger Kalenga-Mpala. 2004. "R&D Activities of the Business Sector in Flanders: Results of the R&D Surveys in the Context of the 3% Target." Brussels: IWT-Studies (46).

Wade, Robert. 1985. "The Market for Public Office: Why the Indian State Is not Better at Development." *World Development* 13:467-497.

Wadhwa, Vivek and Gary Gereffi. 2005. "Framing the Engineering Outsourcing Debate" Duke University: Master of Engineering Management Program.

Wang, Changyong. 2005. "IPR Sails Against Current Stream," *Caijing Magazine*, October 17. Available online at *<http://caijing.hexun.com>*.

Wang, Liwei. 1993. "The Chinese Traditions Inimical to the Patent Law." *Northwestern Journal of International Law and Commerce*, Fall.

The Washington Post. 2006. "Chinese to Develop Sciences, Technology." February 10. P. A16.

Wessner, Charles W. 2005. "Entrepreneurship and the Innovation Ecosystem." In David B. Audretsch, Heike Grimm, and Charles W. Wessner, eds. *Local Heroes in the Global Village: Globalization and the New Entrepreneurship Policies*. New York: Springer.

White House. 2006. "Fact Sheet: United States and India: Strategic Partnership." Press Release. March 2.

World Bank. 2004. *Innovation Systems: World Bank Support of Science and Technology Development*. Vinod Kumar Goel, ed. Washington, D.C.: World Bank.

World Bank. 2006. *The Environment for Innovation in India*. South Asia Private Sector Development and Finance Unit. Washington, D.C.: World Bank.

World Bank. 2006. *World Development Indicators*. Accessed at *<http://devdata.worldbank.org/wdi2006/contents/cover.htm>*.

World Bank International Finance Corporation. 2006. *Doing Business in 2006: Creating Jobs*. Washington, D.C.: International Bank for Reconstruction and Development.

World Trade Organization Working Party on the Accession of China. 2001. *Report of the Working Party on the Accession of China*. WT/MN(01)/3, November 10.

Wu, Ching. 2006. *China to Build 30 New Science and Technology Parks*, SciDev.net, April 19.

Yan, Dai. 2004. "Anti-Monopoly Legislation on the Way." *China Daily*. June 18.

Yang, Lei, ed. 2006. "Chinese WAPI Delegation Quits Prague Meeting." *Xinhua*. June 8. Available at *<http://news.xinhuanet.com>*.

Yoshida, Fujio. 1967. "Preparation of Legal System for Capital Liberalization (Part 3)." Panel Discussion in *Zaikei Shoho*. July 17.

Yukio, Sato. 2001. "The Structure and Perspective of Science and Technology Policy in Japan." In Phillipe Laredo and Phillipe Mustar, eds. *Research and Innovation Policies in the New Global Economy: An International Comparative Analysis*. Cheltenham, UK and Northampton, MA: Edward Elgar, 2001.

Zeigler, Nicholas J. 1997. *Governing Ideas: Strategies for Innovation in France and Germany*. Ithaca, NY and London, UK: Cornell University Press.

Zemin, Jiang, General Secretary of the Communist Party of China Central Committee. 1999. Keynote Speech at the National Technological Innovation Conference, August 23.